FLOWERS OF THE BIBLE
And How to Grow Them

FLOWERS
OF THE
BIBLE
And How to Grow Them

Allan A. Swenson

CITADEL PRESS
Kensington Publishing Corp.
www.kensingtonbooks.com

CITADEL PRESS BOOKS are published by

Kensington Publishing Corp.
850 Third Avenue
New York, NY 10022

The Zones of Plant Hardiness map on page 000 was provided for reprinting by the US Department of Agriculture

All Kensington titles, imprints, and distributed lines are available at special quantity discounts for bulk purchases for sales promotions, premiums, fund-raising, educational, or institutional use. Special book excerpts or customized printings can also be created to fit specific needs. For details, write or phone the office of the Kensington special sales manager: Kensington Publishing Corp., 850 Third Avenue, New York, NY 10022, attn: Special Sales Department, phone 1-800-221-2647.

Illustrations by Peter Jon Swenson

First printing: August 2002

10 9 8 7 6 5 4 3 2 1

Printed in the United States of America

Library of Congress Control Number: 2001099102

ISBN: 0-8065-2314-X

This book is dedicated to the memory of Amy Robinson, whose spiritual insights helped guide me as I researched and wrote this book. Her steadfast faith and good humor in the face of adversity remain a guiding light for all who knew her. To a gentle point of light, my thanks.

Contents

Acknowledgments

Thanks are due to hundreds of people who have devoted countless hours in answering questions, sharing their knowledge of the Scriptures and gardening and otherwise providing worthwhile and helpful guidance in the writing of this book. Special thanks to the Reverend Lamar Robinson for his quiet counsel, Reverend Marsh Hudson-Knapp for his thoughtful suggestions and kind critiques, and my wife, Sheila, whose good sense and editing skills made the text more readable.

Thanks and grateful salutes also to all who have planted and tended Biblical flower gardens and graciously shared their knowledge for this book so others will be encouraged to grow Biblical gardens too. I was pleased to quote many of you in these chapters and especially in Chapter 9, "Biblical Gardens to Visit in the United States," about your glorious gardens. My gratitude especially to Page McMahan, Betty Clement, Joseph Scott, Stan Averbach, and other ardent Biblical gardeners for their help.

To old friends and many new ones I've met of all faiths and denominations—Protestant, Catholic, Jewish—who offered ideas, tips, advice, and down-to-earth Biblical gardening knowhow, my heartfelt thanks. May your gardens grow gloriously and reward you and all who see them with beauty for years to come.

PHOTO AND ART ACKNOWLEDGMENTS

With thanks for their help during research and for providing photos, drawings, and art:

Peter J. Swenson, J. Drayton Hastie, Nogah Hareuveni, Helen Frenkley, Paul Steinfeld, Joseph Scott, Magnolia Gardens, Netherlands Bulb Growers, K. Van Bourgondien, Wayside Gardens, Park Seeds, and the many marvelous Biblical gardens that are identified with their photos.

Allan A. Swenson

Prologue

❧

Let's Grow Together—Biblical Gardens and Life

This is an exciting new millennium, a time of great growing opportunities in our gardens and lives. Gardening has become America's number one hobby. More than 80 million families grow tasty vegetables in their home gardens. Many millions more enjoy the beauty of flowers to grace their home landscapes and indoor living areas too. Today we have a unique new, spiritually uplifting opportunity—Biblical gardens.

The Bible remains America's favorite book. With it and this book also in hand, we can truly dig in to cultivate the blooming beauty of Biblical flowers and grow together better in our love of God and in our appreciation of His blessings. Perhaps we also can cultivate new, worthwhile family relationships in our gardening activities that will sprout, grow, and be rewardingly prolific this year and in the years ahead. My mission, as an avid gardener, an author, and an enthusiastic Christian, is to help families grow together. It is time that we all dig in and with God's help get America growing together.

As you cultivate the beautiful, appealing flowers of the Scriptures, and others that also have their roots in the Holy Land, you will find that through them you can cultivate the best in people and help grow the most important crop that we can grow together, our children and young people everywhere. We can do it with an appreciation of the Bible and the valuable lessons to be learned from it about our role as stewards of the land. We can get gardens and lives growing better with beautiful flowers to satisfy our hearts, minds, and spiritual lives.

We can get growing with the plants of the Bible in our church yard, in the school yard, in our home yards, and even in pots, tubs, and planters on porches, balconies, and rooftops in the cities. We can and should rededicate ourselves to making every square yard more attractive in our neighborhoods. That includes revitalizing and beautifying empty lots in our cities, turning ugly lots into garden spots. As the desert can bloom in the arid Holy Land, city lots can come alive and bloom in urban areas as well. The harvest can be a rich one. Biblical garden blooms to brighten our lives and new friendships in abundance.

As ye sow, so shall ye reap. We can make this a productive year, decade, and millennium of reawakening, of discovery, of a new beginning in our lives. We must do so. America's divorce rate has soared. We are threatened by a frightening crime rate. There is turmoil in the cities. We have poisoned our environment. Some scientists believe that we face a stark and sterile future. That need not be so. We can each do much to make this world a better place, a fruitful place, a more beautiful place. There is really no sterility of ideas available to us. The Bible offers a living daily guide, not just for religious matters, important as they are, but for growing matters as well. It is up to each of us to begin the reblooming of America.

There is much to learn about our land and our gardens and our lives in the Bible. The original lands of milk and honey were just that. People were responsible then. But through the centuries, the land was ravaged by war and erosion and denuded of forests. Today, we must all assume our share of responsibility for making a better world. We must begin again to focus on restoring the goodness to the land and our lives. It has been truly said that mankind must preserve and tend the land and the land will supply bounty in response. It is time we paid closer attention to the teachings laid down in the Bible. As we relearn and apply those lessons, we can and should share them with others. As we do, we can revive and restore another neglected lesson.

It is simply sharing—sharing ideas, sharing ourselves, sharing help to get America and her people growing together again. The pioneering spirit made America the country that it is, a place where opportunities abound for those who have the vision and determination to work together to make good things happen. We often talk of love in this troubled world of ours. Sharing, we should remember, is the true measure of love for our fellow man, one for the other. Are we to be our brother's keeper? Our sister's keeper? We are and must be for those who need our helping hand.

We must look again at our responsibilities, and the challenges, and the opportunities that surround us every day. We are indeed stewards of the land and all that dwell thereon. That is a sweeping responsibility. As youngsters say, it is awesome. Rightfully so. Today, when so much attention is being focused on material things, perhaps using economic terms may drive home some important points. Think of your garden soil and your life as a bank. You can take out only in proportion to what you have put in. If you do not make periodic deposits in your soil bank, you most surely will deplete your soil fertility balance, just as you would deplete your bank account without periodic deposits. To enrich our soil, our good God-given earth, we must use compost, mulch, and fertilizer to build up the right deposits that will lead to rich and abundant harvests and the glorious beauty of abundant flower blooms. This is the first lesson, the most down-to-earth lesson, we can learn. It dates to that admonition in the Bible, to let the land rest. Equally important, if you do not deposit worthwhile ideas into your life bank, you surely will become bankrupt of decent human values.

Gardening begins in the good earth. We must learn more about the earth—how to improve it, how to feed it, and tend it so that it in turn feeds the plants we grow, which in turn feed us—glorious, beautiful flowers as a feast for our eyes and minds, vegetables and fruit as feasts for our bodies. There are other lessons of life and living to be learned from our gardens. It is time that we follow another passage or two. Seek and Ye Shall Find. Ask and It Shall Be Given. We can seek, seek an appreciation and understanding of these beautiful passages from the Bible and the guidance they provide for helping us improve our growing world. Today we can begin anew to grow the flowers of the Bible, to trace their roots to the Holy Land, to follow good growing techniques as they were handed down to us from Biblical days, with the lessons others have added to our horticultural knowledge through the centuries. We should pause at times to think and reflect and then resolve to get ourselves growing better with our families and our church and community families for a better world.

As we read again from Scriptures, there is a time to sow and a time to reap. Now is the time to sow, to sow seeds for plants, to sow ideas for Biblical gardens and the lessons they can teach us for our daily lives. It is the time to sow the seeds of new friendships, reaching out to cultivate others in our neighborhoods, our communities, our cities, and our country. Families that sow together can truly grow together. And that is good. As we work with our young people and they with us, we can see the miracles of God's world sprout and grow and bloom before our eyes. These gardens provide growing experiences for all of us, young and old.

With flowers of the Bible, you add another dimension to the usual garden's growing experiences. You expand the horizons of discovery, tracing those plant roots back to the Scriptures and Holy Land and refreshing your mind again with the meaning of gardens and life and living. The world is a growing experience. Today, there is an increasing need to put our faith to work again. One rewarding way is with flowers of the Bible and Holy Land that will sprout, take root, grow, and rekindle for us all the messages given to us so long ago. As we sow better growing ideas and share our knowledge and lend helping hands to others, so shall we reap. We shall indeed harvest an abundance of beauty from our Biblical flower gardens; gain new growing horizons with children, family, and friends; and also be an important contributor to making our shared world a much better one.

That is the message I wish to plant with you in this book. Learn about the flowers of the Bible and the beauty that they give to us all. Read the Bible with new insight. Then, dig in, and plant the seeds and bulbs when and where and as you can. Invite friends and neighbors to garden with you and help them as they expand their gardening and living horizons. Cultivate your gardens and your family and your friends and your neighbors, and together we can all help get America and our world growing better again. It will be a most beautiful sight!

Allan A. Swenson
Windrows Farm
Kennebunk, Maine

Chapter One

✿

Biblical Growing Horizons

Flowers of the Bible are alive and thriving today, perhaps much closer to your home than you may realize. You probably pass some of them on your way to work or have a few in your home garden already. Anemones, crocuses, hyacinths, irises, lilies, and tulips all have their historic roots in the soil of the Holy Land and in the Scriptures too.

Through many centuries, Biblical flowers were faithfully grown in the gardens of religious orders, as I've seen in person at monasteries and convents during research for this book. Many more have been growing gracefully in local churchyards, and you'll find many Biblical gardens that you can visit in Chapter 9.

Today, as a new and welcome feeling of spirituality grows among us all across America, focusing on creating Biblical gardens seems especially appropriate. You, too, can enjoy the blooming beauty and fragrance of many of these plants that are mentioned in the Scriptures. Actually, the Bible abounds in references to plants, as all who have read it realize. From earliest descriptions of the Garden of Eden through books of the Old and New Testament, we can read passages depicting flowers, trees, fruits, foods, and herbs. You'll find those Scriptural references with the appropriate flowers in this book.

We North Americans and Europeans, of course, accept our

four-season year as normal. However, the growing season is quite different in the Holy Land, where there are only two seasons—a hot, dry, dormant season and a rainy growing season. The Holy Land has always been a crossroads of the world, linking Europe, Asia, and Africa. Caravans traveled through that area in antiquity with goods for trade. Plants and seeds were included. Over the years, pilgrims and travelers have visited and transplanted plants of the Bible to the far corners of the globe. People have always been attracted to plants and naturally want to take some of those that appeal to them back to their homelands to grow in their own gardens.

At other times, war brought people to that part of the world. Plants originally native to India, Africa, China, and much of Europe were carried on caravans across hundreds and hundreds of miles. Orchards and vineyards, grain fields, and vegetable gardens were planted for food. As people moved and warred and traded, it was also natural that this interchange of plant varieties would take place. Flowers were brought to new areas as well. You'll learn some of their history in the chapters about them.

As you read the many references to plants that are found in the Holy Land, you may wonder whether you can grow them yourself. The answer is a resounding and emphatic *yes!* I've spent a decade doing the research to help you grow Biblical flowers, gloriously, bountifully, and fragrantly too. During the past decades, millions of people like you have discovered the pleasures of home gardening. Today gardening is America's most popular family hobby. In response to recent Gallup polls, more than nearly 100 million people report that they garden in one way or another. With this increased popularity, more people have begun focusing on special types of plants that would give their gardens a different and distinctive look. Along with the new millennium has also come a growing sense of spirituality in America. Church leaders I've interviewed often commented on this new feeling that is growing among their communities and in their congregations and parishes.

After the terrible terrorist attack of September 11, 2001, people of many different faiths and denominations came together in a demonstration of their faith in God and their love of country. Today we are seeing a greater understanding of different cultures and also an appreciation of the need to work together in our global society. Among church and synagogue members there is renewed attention to the Bible and among gardeners to Biblical plants, a welcome sign indeed. I discovered that as more people looked for special plants to make their gardens more distinctive and attractive, they too realized that Biblical plants have special meaning dating back thousands of years to the Holy Land. That growing interest led me to research new gardening trends, which led naturally to this book.

HISTORIC PERSPECTIVE CAN HELP

The Romans traded throughout their Empire, which stretched through the Middle East, into Africa, and across much of Europe. During the Crusades plants were taken home for use in the countries from which the Crusaders, pilgrims, and also merchants had come. Today many of the wild and domesticated flowers that trace their heritage throughout pages of the Bible have numerous hybrid relatives around the world. Enthusiastic gardeners today are turning their attention to plants with historic importance and spiritual significance.

Another trend has been helpful. Heirloom plants, some of those purebred original flowers, are being rediscovered and brought back to cultivation. Several mail-order firms, including Burpee, Seeds of Distinction, Select Seeds, Heirloom Roses, and other firms, are finding and offering marvelous old-time favorites. In addition, plant breeders are creating dazzling new hybrid flowers that you can find on chain store seed racks or order by mail. A list of many reliable firms that offer free catalogs with Biblical plants is included in Chapters 11 and 12.

Some Biblical plants and relatives already exist in the United States because early settlers brought seeds and plants with them as they came to the New World. For those among us who also enjoy the beauty of wildflowers, the Bible mentions many that grow in the Holy Land, and their wild relatives are found in many other parts of the world. Daisies are one classic example. Roses are another.

BIBLE TRANSLATIONS MAY NOT IDENTIFY PLANTS RIGHT

The Old Testament, or Hebrew Bible, originated in the form of songs, ballads, and poetry handed down from generation to generation by word of mouth. Today we call that oral history. The moral and religious messages were of far greater importance than accurate plant identifications. The Bible was never intended by its writers, translators, or religious leaders to serve as a botanical or even natural science text or reference. Among Biblical scholars and botanists alike, today as in the past, debate continues concerning the true identity of certain specific flowers, fruits, trees, and other plants mentioned in Scriptural passages. Considering the number of different translations of the Bible, especially from the ancient Greek and Hebrew into various modern languages, confusion is understandable.

The widely accepted Authorized Version of the King James Bible contains some misidentification of plants between the Biblical ones and common English ones. For example, aspens have been called mulberries and mulberries called sycamores. Add to that confusion the different translations in the New International Version, The New English Bible, Today's English Version, the Goodspeed, Jastrow and Moffatt, and other versions, and we gardeners rightfully scratch our heads in wonder.

Goodspeed, Jastrow, and others have variously identified the "lily of the valley" in the Song of Solomon as an anemone, hyacinth, narcissus, or lily. Artists during the Middle Ages and Renaissance often depicted the Madonna lily with the Annunciation and Resurrection. Many botanists concur that based on botanical and ecology factors, the lily of the fields, or lily of the valley, are more likely references to the Palestine anemone.

Despite these debates and the lack of exact definition of some plants, most have actually been rather clearly and accurately identified. I've spent more than a decade doing research and interviews with experts for my earlier Biblical plant books and this one. This has been possible by careful comparison of the old Scriptures and translations with the plants that are native today in the climates and soils of the Holy Land. By studying botanical nomenclature, tracing plant families and plant structure and their native habitats, today we can be reasonably specific in identifying the majority of flowers of the Scriptures.

Beyond those plants that we can rather accurately identify as being in the Bible, there are dozens of others that grow wild and naturally in the Holy Land today as they have for millennia. Although it may not be possible to clearly identify them from specific Scriptural passages, they are obviously plants of the Holy Land and deserve a place in this book and Biblical gardens. If specific species and varieties are not conveniently available, they can be represented by related look-alikes from their genus.

ECOLOGY AND SEASONS OF THE HOLY LAND

Let's go into the seasons issue in a bit more detail to better understand these Biblical plants. We who live in North America and Europe naturally plan our gardens based on four seasons. As residents of Florida, Southern California, and areas along the Gulf Coast know, growing seasons can be quite different in those parts of the United States, especially if they have lived and gardened in Northern states. In the Holy Land there are two seasons, the dry dormant time and the rainy growing season, which I'll describe in more detail later.

These two seasons are rather different from what we might expect. The "winter" is actually the growing season because it is the time of rain when plants are given the moisture they need to burst forth into glorious bloom. It is a time when seeds sprout and rush to produce plants, which in turn bloom, set seeds, and die as annual plants do. Even perennials must grow, bloom and mature in this short time when moisture arrives.

The Holy Land, at the crossroads of three continents, Europe, Asia Minor, and Africa, has flora that are unique in the world. The Holy Land was part of the fertile crescent, which we recall from our history classes was one of the earliest centers of civilization. As few as 15,000 years ago, scientists estimate, the Holy Land was indeed a literal Garden of Eden. Lush vegetation abounded and the fertile valleys, wooded mountains, and tropical oases provided abundant sustenance for those who lived there. Archeological digs have provided ample evidence of the abundant and wide range of vegetation that had existed there. When Moses led the children of Israel in their flight from slavery in Egypt through the wilderness, their goal was the Promised Land. Even then the land had an abundance of plants.

Time and events, unfortunately, have altered the land. Despite humanity's depredations against neighbors in wars and misuse of the land itself by overcultivating it, denuding forests of trees, overgrazing by livestock, plants have more resilience than we might expect. That is most fortunate for the plants and mankind as well. Despite what people have done to destroy their environment through the ages, plants seem to have a way to survive and make their comeback.

The Scriptures also give us some guidance as we embark on our own Biblical gardens today. Too often we do not pay sufficient attention to protecting and preserving our environment. Fortunately, with recycling, energy conservation, and natural gardening without harmful chemicals, this era seems to have come alive with much greater focus and appreciation for our shared environment. That is a good sign for all. It may prove worthwhile to note that conservation and care of the environment were early topics in the Bible. If humans had focused on this wisdom in the Scriptures, perhaps we would not have seen the destruction of the land in so many parts of the world.

By removing trees and other large plants, continual planting of one crop only, and neglecting to replace soil nutrients taken out of the land by plants, we have caused serious depletion of natural resources. Proof of these tragic errors includes the dust bowls created in our own Midwest during the last century. Wars and unrestricted farming have also savaged and ravaged the Holy Land. Few of the forests once blanketing the land in Biblical times now remain or even many examples of the Cedars of Lebanon of Biblical fame.

In Exodus 23:10–11 we can find good advice about conservation and proper land use. "And six years thou shalt sow thy land, and shalt gather in the fruits thereof: But the seventh year thou shalt let it rest and lie still . . ." From Leviticus 24:3–5 we also receive an admonition to allow land a time of rest. "Six years thou shalt sow they field, and six years though shalt prune thy vineyard, and gather in the fruit thereof; But in the seventh year shall be a sabbath of rest unto the land, a sabbath for the Lord: thou shalt neither sow thy field, nor prune thy vineyard. That which groweth of its own accord of thy harvest thou shalt not reap, neither gather the grapes of thy vine undressed: for it is a year of rest unto the land."

From Deuteronomy 8:7–8 we can visualize what the land of the Bible was: "For the Lord they God bringeth thee into a good land, a land of brooks of water, of fountains and depths that spring out of valleys and hills; A land of wheat , and barley, and vines, and fig trees, and pomegranates; a land of oil olive, and honey." Also from Deuteronomy 8:15–16, we learn how other parts of the land and its adjacent areas must have been in great contrast: "Who led thee through that great and terrible wilderness, wherein were fiery serpents, and scorpions, and drought, where there was no water; who brought thee forth water out of the block of lime; Who fed thee in the wilderness with manna, which thy fathers knew not, that he might humble thee, and that he might provide thee, to do thee good at thy latter end."

These Scriptural descriptions well describe the land of olden days. There were indeed deserts, which had to be crossed and they exist there today, inhospitable to most plants except those that are adapted to harsh conditions of baking sun and sparse rainfall. Yet, even in deserts there is a time when plant life bursts forth. Anyone who has seen the colorful blush and sudden flush of spring in America's southwestern deserts can realize what a few drops of rain in spring can mean. Seeds await that time to sprout, rush to bloom, and scatter their seeds, which in turn await the rain of the following year. Such is life in desert regions of the world.

CLIMATE ZONES OF THE HOLY LAND

There are three basic climatic zones in the Holy Land: the Mediterranean Zone, the Irano-Turanian Steppe Zone, and the Desert Zone. Each has its distinctive features, but these zones are not as clearly defined as one might expect. In fact, adjacent valleys may reflect the plant life of the Mediterranean Zone while the mountains between support the vegetation of the steppe. There are places in which the transition of one zone of lush, verdant growth changes gradually to a less productive area. Sometimes, however, the change is abrupt with plant life changing dramatically within less than a mile from verdant, blooming beautiful scenes to virtual desert conditions. There is truly much diversity and contrast in so small an area.

The Mediterranean Zone is characterized by a dry, not overly hot summer and winter months of rainfall. The Irano-Turanian Steppe Zone has long periods of dryness and a scant ten inches of rainfall annually. The Desert Zone is hot and arid with negligible rainfall. You can get a good picture of these zones from modern atlases. My *Rand McNally Millennium World Atlas* provides some fine maps. Best of all is the *Westminster Historical Atlas to the Bible*. In it you can find maps that depict the Holy Land during the Roman era, the Period of the Judges, various parts of Palestine in Biblical times, plus much good historic information about the land and the people who have inhabited it for the past millennia.

Naturally, as expected, the Mediterranean Zone provides the most favorable conditions for plant growth. Rainfall ranges from 20 to 40 inches per year and occurs during the winter months. Temperatures rarely fall below freezing in this zone but the dry period of summer serves as the dormant or resting period for most plant life.

The rain that begins in late September to early October marks the beginning of the growing time. Scattered showers appear, developing into heavier rains during December and January and slackening during April. This warm, rainy winter season is the peak growing period for most crops and flowers. However, as horticulturists explain, summer dews that collect during cool nights do provide some moisture for plants. This can be compared to 10 inches of rainfall during the dry-appearing summer season.

The type of soil, amount of rainfall, and range of temperatures are key factors that determine the type of plant growth that appears. Both coniferous forests of pine and deciduous oak trees are present, along with terebinth and carob trees, which are two native trees. A variety of flowers grow wild in this area, gracing the land with color. You'll find details about flowers and where they grow and how to grow them in Chapters 2 and 3.

The Irano-Turanian Steppe Zone is a dividing zone separating the Mediterranean Zone from the Desert Zone. Please don't confuse the term "steppe" here with the frigid steppes of arctic and near-arctic areas. The steppes of the Holy Land are distinguished by their dryness rather than cold. Rainfall is scarce, seldom reaching 10 inches annually, and only during the winter season from December to March. Other months are almost totally dry.

This limited moisture factor, presence or lack of water, is the key that influences plant life. Few plants can tolerate such difficult growing conditions. However, annuals that can sprout, bloom, and reseed themselves in a short period do exist and seem to have flourished through the centuries. I've included in Chapter 3 some that would grow here.

Among the different flowers of this area, bulbous plants from crocuses to tulips do reasonably well; they store nutrients in their enlarged roots or bulbs. Trees and bushes also have solved the climate problem. They shed their leaves in dry weather much as deciduous trees in temperate climates do in the fall when cold weather arrives.

Flowers and plants in the Steppe Zone, which is a much smaller area compared to the other zones, as you will see when you look at maps in atlases, burst into bloom at their appointed time. As those who have visited the Holy Land can attest, in winter and spring the landscape changes dramatically from its drab, dry-looking appearance to incredible blooming beauty. Bulbous flowers send up their preformed leaves and blooms once the touch of rain gives them the signal. Wild annual flowers sprout and surge quickly. It is a remarkable time and sight, much as America's great southwestern deserts look as they come alive for a brief but dramatically colorful profusion of bloom, which then as swiftly is gone. As the brief growing period expires in late spring, the weather forces the land back to its brownish drabness. Thus it is with semiarid areas around the globe, causing those who live in them to cherish that brief span of bloom with great appreciation.

The third Holy Land area is the Desert Zone. For most of the year it is an area devoid of apparent plant life. During the scorching heat of summer, no plants appear, but the roots and seeds of hardy plants lie secretly awaiting a brief chance at life and beauty.

This Desert Zone has climatic conditions similar to those of the Sahara-Arabian. In the Holy Land, the desert actually is bounded on the west by the Sinai, which includes most of the Negev Desert. To the south it circles around the Mediterranean and Steppe areas and links into the expanses of the vast Arabian Desert.

As you consult the Bible about plants, you also may be surprised to realize that there is really no mention of four seasons in it. In the Song of Solomon, 2:11–13, you will find this passage: "For lo, the winter is past, the rain is over and gone; The flowers appear on the earth; the time of the singing of birds is come, and the voice of the turtle is heard in our land; The fig tree putteth forth her green figs, and the vines with the tender grape give a good smell. Arise, my love, my fair one, and come away."

You may search further, even through the Talmud, the Jewish Holy Book, but you will find only reference to the two seasons. Those are the growing period through harvest and the time of the resting of the land. In effect you will learn from both the Talmud and the Bible that there are the days of sun and the days of rain. That factor is the controlling one for plant and flower growth there.

Search other parts of the Mediterranean Sea basin, Greece, Turkey, Italy, Egypt, and North African nations for climate variations and you will find that other countries have basically the same two-season year. But the Holy Land is distinctive, as comparatively it has the least rainfall of any other countries in the region.

BLOOMING TIMES TO VISIT AND TOUR

For those of you who may someday plan a visit or pilgrimage to the Holy Land, perhaps an overview of the natural cycles of flora will be worthwhile. As a gardener, I think in terms of the growing year as I plan trips and vacations and this might be pertinent to you as well. If you plan with an eye toward growing seasons, you will be able to see gardens at their peak, for photos, comparisons of plants, and the general beauty they provide at their zenith.

In the Holy Land, the growing season begins in October as the first rains arrive, plants are awakened, and regrowth is stimulated. It is most logical that the Jewish New Year, Rosh Hashanah, is in October. When October rains fall on mountain slopes, meadows, and fields, the flowers of the well-designed bulbous plants begin their blooming cycle. Crocus and narcissus rise and bloom. The whitish petals with orange crown in the center are a dramatic, welcome sight.

Cyclamens begin, spreading their heart-shaped leaves among rocks and sending their single, crownlike blooms up on long stems. This curious flower blooms early and lasts longer than most others, with final blooms in May in the Upper Galilee area. With the encouraging rains, other flowers, such as tulips, raise their showy blooms, and anemones get ready for their flowering. Anemones are found everywhere from the Negev Desert to the Galilee area. Later in winter hundreds of red anemones blaze across the fields.

By November, rains will have worked their wonders, awakening the other dormant plants. Narcissi are in their glory and crocuses dot the countryside. Cyclamen are in dazzling beauty while farmers are busy in their fields with their crops as they have responded for centuries to that growing cycle.

In December, much of the land will have turned verdant with increased rains. Berries and fruits including oranges, lemons, and grapefruit will begin maturing and ripen. During January, the anemones continue their colorful display, showing their bright reds and purples to the landscape. Now irises will begin their blooming beauty alongside streams and waterways.

By February, more trees blossom along with iris, one of the striking flowers in the hills around Nazareth and in the northern Negev area. Botanists explain that there are several distinctive irises native to the Holy Land, but the typical Flag iris seems to be most common with its blue to purple hues. The constant anemones and hyacinths add color to the landscape and fragrance to the air. It is indeed a sweet time of bloom. For residents and those people who have visited, this is one of the most beautiful times in the Holy Land.

In March, the large iris and tulip, lupine, and other lovely plants carpet the countryside. Gardens, so lovingly planted and tended as rains of winter fell, are reaching their prime in March. By May, poppies and thistles, lilies and flowers that prosper with the warmth of summer begin their bountiful displays of beauty. In June, depending on the zone, late summer flowers appear. It is the time when vegetable and fruit harvests begin from family plots and commercial orchards and farms alike.

In late July and August, the warm, drying winds will regain dominance over the land. Plants will wither and the land returns to its natural dormancy without the blessing of rains to support plant growth. In September, many plants near their end. Seeds are set and dropped. New blooms and nutrients for next year's plant growth are safely deposited in the bulbs, corms, tubers, and rhizomes of flower plants. That is the natural cycle, blooming and then storing nutrients in the respective root areas for the coming year's plant.

The plants lie waiting, as perennial flowers do in North

America and Europe during the northern climate winter season. There, the cold weather is the factor that induces dormancy in flowers and plants each fall. Snow may blanket the ground up to several feet deep, yet tucked safely below the snow, the bulbous flowers and roots of perennial plants lie safely protected. In the Holy Land, the roots and bulbs lie sleeping through the dry time.

With the approach of October, the winter rains traditionally appear again as they have for centuries and aeons before. With the rain, the growing cycle begins again. The land will once more sprout into greenery, bloom, and bear. From this land have sprung the plants of the Scriptures.

Chapter Two

✿

Flowers of Biblical Scriptures

The Holy Land, Asia Minor, and the countries bordering the Mediterranean Sea are the sources of the classic bulbous flowering plants so popular in America and Europe today. From these regions have come our crocuses and cyclamens, hyacinths and irises, lilies and tulips. These flowers developed here under the hot dry season, which forced them into dormancy, and then the brief moist period, which forced them into bloom.

The term "bulb" is a horticultural term that is usually interpreted rather broadly. It includes the true bulbs, which have food storage parts within the bulb. For example, cut a tulip or narcissus bulb in half and you will find the makings of a complete plant: leaves, stem, and flowers with its food stored within. Today, the word "bulb" also has come to mean similar bulbous roots which are called corms, rhizomes, and tubers by botanists. These plants are ideally suited for survival and adaptation.

Because many of the Biblical flowers grow from bulbs, other characteristics of the true bulb include the central growth bud, which is sheathed in graded layers of modified white leaves. These are known as scales on plants such as lilies, narcissi, and tulips.

Corms are slightly different. They are bulb-like but consist of a fleshier base of the stem which stores up needed nutrients made and deposited by the plant during the previous year's growing season. As with bulbs, that food reserve is stored in the scales that surround the tiny growth bud from which the new flower plant will form. A crocus grows from a corm as does the gladiolus.

Some bulbous plants have tubers. These can best be described as thickened underground stems with rough, leathery coating. Cyclamens and begonias grow from tubers. Still another type of "bulb" is called a rhizome. This is thinner and more elongated than the tuber's underground root. It is formed by the stem of the plant and in fact is actually an underground stem or root-stock that bears buds. The iris is typical of plants that grow from rhizomes.

Let's now go to the Bible, where we first learn about Biblical plants. In the beginning, God created the earth and the waters; and then He created the plants on the third day. As we read in Genesis 1:11–12:

> Let the earth bring forth grass, the herb yielding seed, and the fruit tree yielding fruit after his kind, whose seed is in itself, upon the earth: and it was so. And the earth brought forth grass, and the herb yielding seed after his kind, and the tree yielding fruit, whose seed was in itself, after his kind: and God saw that it was good.—King James Version

We hear God calling into existence an array of beautiful new life. God seemed to delight in gardens, as we do. That is only natural. As we read further in Genesis 2:15, we discover that God's primary purpose for creating people was to be gardeners of the good earth. Consider, for example, that the Bible is a book of gardens from Eden to Gethsemane. King Solomon's garden contained many flowering and aromatic plants, including lilies and crocuses.

In Biblical times average people could not afford such a luxurious garden as King Solomon's, but they lived in a land where the hillsides and valleys were profuse with anemones, narcissi, hyacinths, tulips, and many other flowering delights. A garden is essential to a person's life, and in death one is often buried in a cemetery garden setting. Gardens were gathering places for way-farers through the centuries. Jesus and his disciples often went to gardens to rest, contemplate, and pray. It was in the garden of Gethsemane that Jesus made his great decision and it was in the garden of Joseph of Arimathea that the body of Jesus was laid to rest.

During my years of research and writing my two earlier books, *Your Biblical Garden,* published by Doubleday in 1981, and its sequel, *Plants of the Bible and How to Grow Them,* pub-lished by Citadel and Birch Lane Press in 1994, I carefully fo-cused on the flowers that most Biblical and botanical scholars have identified as being mentioned in the Scriptures. There has been much honest debate over the aeons, for good reason. There still is. Here is my list for this book. I have thought carefully about which flowers I should include as truly representative of those mentioned in the Scriptures and those that also grow in the Holy Land.

SELECTING BIBLICAL FLOWERS

I have based my list on several key considerations:

First, I have included those plants that Biblical and botanical scholars generally agree are mentioned in the Scriptures. You'll find those in this chapter.

Second, I have listed those flowers that grow in the Holy Land, but may not be specifically identified in Scriptural pas-sages. You'll find those in the next chapter. These would be the most commonly found wildflowers, represented today by simi-lar look-alikes in various floral families.

Third, all the flowers in this book can be traced to historic writings and are also found today growing in the Holy Land, both wild and as domesticated relatives.

Fourth, the flowers must have eye appeal and be beautiful. Naturally beauty is in the eye of the beholder. Some may view mustard as beautiful, but even with its historic roots, I have not included that or some other "flowers of the field." Frankly, there are only so many pages in a book and I had to make logical decisions about which would be the best flowers to include.

Fifth, the flowers must be reasonably available, either the actual flowers or closely related family members that can be purchased locally in garden centers, in stores, or from reputable mail-order firms.

Finally, the flowers must be those that can be grown in most parts of the United States. I have included some that are sensitive to cold, but they can be used outdoors and brought indoors to enjoy and protect for the winter.

With those considerations, I wish you all great growing. My key list is on the next page. As mentioned above, it includes both flowers with Scriptural references plus native flowers of the Holy Land that have no specific Scriptural identification but have their roots there from Biblical times and grow wild there naturally today.

You'll most likely enjoy growing your own Biblical garden when you are well versed in the facts of these flowers' history, habitats, and seasons and the ecology of the Holy Land. There are many fascinating facts and information in that later chapter, which I included for those who want a thorough resource about both Biblical plants and their native growing habitat.

When you understand the Holy Land seasons, environment, and ecology, you will be better informed, more successful, and more capable of guiding others to their own successful Biblical gardens. However, if you are anxious to dig in and get growing, you may elect to focus on the flowers themselves and the growing tips and ideas. Many gardeners prefer to dig in and get growing with hands-on know-how! These next two chapters feature knowledge of plants themselves and provide tips for growing them successfully.

DETERMINING WHICH ARE SCRIPTURAL PLANTS

When the earliest translations of the Bible were made, there were honest differences of opinion about what specific words meant. To linguists, ancient Hebrew and Greek words could be

Master List of Biblical Flowers in the Scriptures and
Flowers of the Holy Land

Flower Name	Scriptual Reference	Latin Name
Anemone	Matthew 6:28	*Anemone coronaria*
Buttercup	Psalms 4:5	*Ranunculus*
Saffron Crocus	Isaiah 35:1, Song of Solomon 4:13	*Crocus sativum*
Chrysanthemum	Various	*Chrysanthemum coronarium L.*
Cyclamen	Song of Solomon	*Cyclamen persicum*
Daffodil, Trumpet Narcissus	Matthew 6:29, Isaiah 35:1	*Narcissus sp. or N. tazetta*
Daisy, Shasta	1 Kings 6:29, Psalm 45	*Chrysanthemum sp.*
Dandelion	Exodus 12:8	*Taraxacum officinale*
Dove's Dung, i.e., Star-of-Bethlehem	II Kings 6:25	*Ornithogalum umbellatum*
Flax, Blue Perennial	John 19:40, Proverbs 31:13	*Linum perenne*
Hibiscus	James 1:9–10, Job 30:4	*Hisbiscus syriacus*
Hollyhock	Job 6:6–7	*Alcea setosa*
Hyacinth	Song of Solomon 6:2–3	*Hyacinthus orientalis*
Hyacinth, Grape	Song of Solomon 6:3	*Muscari sp.*
Iris, Blue Flag	Hosea 14:5, Song of Solomon	*Iris versicolor*
Iris, Purple	Hosea 14:5	*Iris astropupurea*
Iris, Yellow Flag	Hosea 14:5	*Iris pseudacorus*
Larkspur, Delphinium	Luke 12:27, Song of Solomon 6:24	*Delphinium ajacis*
Lavender	Song of Solomon 6:24	*Lavendula stoechas*
Lenton Rose	Song of Solomon 6:24	*Helleborus*
Lily, Madonna	Song of Songs 2:1–2	*Lilium candidum*
Lily, Water	1 Kings 7:19, 1 Kings 7:22	*Nymphaea lotus—white* *Nyphaea caerulea—blue*

Flower Name	Scriptual Reference	Latin Name
Lupine	James 1:9–10	*Lupinus augustifolius*
Loosestrife	Song of Solomon 1:14	*Lythrum salicaria*
Mallow	Job 6:6	*Malva moschata*
Narcissus	Matthew 6:30, Song of Solomon	*Narcissus tazetta L.*
Mum, Hardy	James 1:9–10	*Chrysanthemum sp.*
Poppy, Common	Isaiah 40:6, Jeremiah 8:14	*Papaver rhoeas*
Poppy, Icelandic	Isaiah 40:8	*Papaver alpinum*
Poppy, Oriental	Isaiah 40:6	*Papaver orientale*
Red Buttercup	1 Peter 1:24	*Ranunculus, asiaticus*
Rockrose	Isaiah 35:1	*Cistus sp., Cistus creticus L.*
Rose, Rugosa	Isaiah 35:1, Wisdom 24:14	*Rosa rugosa*
Phoenicia Rose	Eccesiasticus 24:14	*Rosa phoenicia*
Salvia-Sage	Exodus 25:31–33	*Salvia officinalis*
Globe Thistle	Judges 8:16	*Echinops viscosus DC.*
Tulip, Red, Sharon	Song of Solomon 2:12	*Tulipa sp.*
Violet	James 1:10, Song of Songs	*Viola ordorata*

interpreted to mean several different things. The earliest translators of the Bible never saw themselves as "linguists." Further problems arose whenever people translated a Biblical text from one language to another. This translation problem was compounded by the fact that there was no botanical nomenclature when the Scriptures were first translated. Furthermore, although some translators relied on their best knowledge, most translators were and even today are not botanists familiar with the history and categories of plants native to Palestine. As I wrote my earlier books and continued detailed research for this one, I have tried to consult with as many Biblical and botanical scholars as possible. From that perspective, this chapter includes the flowers that most authorities agree are the likely flowers mentioned in the Scriptures. You'll also find planting and care tips about them. Happily, all are bulbous and perennials. Once you prepare your garden ground well and plant them, you'll be rewarded with their blooming abundance for years to come.

There are many other flowers that could be identified as having been mentioned in the Scriptures, but in more generic terms. They do indeed grow in the Holy Land and their roots have been there for centuries too. For this reason, many who have planted Biblical gardens have included those other flowers in their plantings. That seems especially appropriate to present the widest diversity of Biblical flowers and other flora in this resource book.

In the following chapter, I have included many other equally beautiful and appealing flowers that are deeply rooted in the Holy Land. You may wish to begin your gardens with the plants in this chapter, and then expand your growing horizons with flowers from the following chapter. I have also provided the probable or possible Scriptural references to those other flowers

Flowers With Traceable Scriptural References

Biblical Flower	Scriptural Passage	Latin Name
Anemone	Matthew 6:28	Anemone coronaria
Saffron Crocus	Isaiah 35:1, Song of Solomon 4:13	Crocus sativum
Cyclamen	Matthew 6:29, Isaiah 35:1	Cyclamen persicum
Daffodil, Trumpet Narcissus	Matthew 6:29, Isaiah 35:1	Narcissus sp. or N. tazetta
Hyacinth	Song of Solomon 6:2–3	Hyacinthus orientalis
Iris, Blue Flag	Hosea 14:5	Iris versicolor
Iris, Yellow Flag	Hosea 14:5	Iris pseudacorus
Lily, Madonna	Song of Songs 2:1–2	Lilium Candidum
Narcissis, Daffodil	Matthew 6:30,	Narcissus tazetta L.
Star-of-Bethlehem	II Kings 6:25	Ornithogalum umbellatum
Tulip, Red, Sharon	Song of Solomon 2:12	Tulipa sp.

in the next chapter for reference. In all such research, I have relied on the knowledge of many experts in their fields to allow the broadest interpretations. That has advantages because a more open, diverse list permits a wider choice of flowers.

My theme for this book is down-to-earth and, I trust, uplifting: Let's Grow Together. As I traveled to do research, interview, and learn more about Biblical plants and gardening, I have been impressed with the willingness of people to share their knowledge. Today, there does indeed seem to be a coming together of people, a new sense of spirituality and appreciation of the need to grow together in many important ways. I have sensed this feeling in visits with people of all faiths and denominations.

We may each hold true to our personal faiths and beliefs, but there seems a new willingness to accept each other's differences as we understand how much we have in common and to share. This is a welcome feeling well worth cultivating in our gardens and lives. Here are the flowers mentioned in Holy Scripture and how to grow each of them successfully. You'll note that the lily is listed first although the rest of the plants are in alphabetical order. I did that on purpose because there has been so much debate about the true identification of the lily. It seemed best to illustrate how Biblical translations and botanical identifications have sometimes created misunderstanding about which plant is really meant in certain Scriptures.

LOVELIEST LILIES

"I am the rose of Sharon, a lily of the valleys. As a lily among thorns, so is my love among the daughters."
—Song of Solomon 2:1–2

"I will be as the dew to Israel; he shall blossom as the lily, he shall strike root as the poplar."—Hosea 14:5

"The desert shall rejoice and blossom; like the crocus it shall blossom abundantly, and rejoice with joy and singing."—Isaiah 35:1–2

"His cheeks are as a bed of spices, as sweet flowers: his lips like lilies, dropping sweet smelling myrrh."—Song of Solomon 5:13

Lilies are the most commonly named flowers in the Bible. Yet, there still is debate about what actual flower is meant. Different translations vary considerably too. As we consider this truly Biblical flower for our gardens, it is pleasant to read about it in the Scriptures. Although much discussion through the ages has centered on the identification of the plant or plants mentioned in the scriptures as lily or lilies, this passage seems to eliminate the Madonna lily, *Ilium candidum*. Through the centuries, particularly through the Renaissance, attention was focused on the lily as a pure white flower. Therefore, it is somewhat confusing to expect that in the Song of Solomon, lips would be described as pure white.

Considering the context of this verse, we must think in terms of flowers that have a red color. Some scholars have suggested the scarlet lily, *Ilium chalcedonicum*, and the Martagon lily, *Ilium martagon*, for the honor. This passage seems to warrant a plant of exceptional beauty. However, the Martagon lily is rare in the Holy Land, and according to botanists, it was never native there.

In the Goodspeed version of the Bible, the word "hyacinth" replaces lily in this passage. The actual existence of the Madonna lily, *Lilium candidum*, in the Holy Land has been a topic of controversy for centuries. The debate still rages on, with those pointing to the few rare Madonna lilies that still can be find at higher altitudes in some parts of Israel today, to those who fervently believe that other lilies were intended because they grow more abundantly and widely today as they did millennia ago.

It is true that famous artists depicted Mary with a Madonna lily in her hand. This lily has been a favorite at Easter for centuries. The pure white lily has been a symbol of innocence, virginity and purity, and the resurrection. It is found planted in many church graveyards where the climate allows them to survive. Every Easter, the white lily adorns the sanctuaries and altars of churches nationwide.

Some botanists who have meticulously studied plants mentioned in Scripture do believe the Madonna lily is one of the lilies found in Solomon's famous garden, as noted in Song of Solomon 6:2 and also was the lily mentioned in Hosea 14:5. It is a fact that early in the twentieth century, botanists did find colonies of this plant growing in northern Palestine. Neot Kedumim authorities note that this lily really grows only in the Mountain Carmel area.

One other key historic point has been a part of the debate whether the pure white Madonna lily is in fact the lily of the Scriptures.

A papal edict issued in 1618 laid down stringent rules concerning the appropriate treatment of certain sacred subjects in art. In this edict, the necessity of showing the white lily in paintings of the Annunciation was emphasized. For centuries, famous artists depicted only a white lily in their Annunciation paintings. That fact, some scholars point out, may have been more influential in identifying the lily than any botanical or natural ecology facts.

The Biblical Hebrew term "shoshan" or "shushan" is a white lily, according to some linguists. Reportedly the white lily that grows in Galilee and on Mountain Carmel was very common years ago, but it is rather rare today. There are scholars, Biblical and botanical alike, who focus on the Scarlet Turk's Cap lily, Lillim chalcedonicum, as the true lily of Scriptures. The maiden who compares the lips of her beloved shepherd to a lily would hardly exclaim about white lips. They say that passage proves the intended lily was red. Others part company and focus on the red anemone. You can make up your mind as you, too, read Solomon 6:23 and 5:13.

Perhaps the Scarlet Turk's Cap lily, is a likely contender. This lily grows to 4 feet tall with brilliant scarlet flowers. You may find appropriate Turk's Cap lilies in your search through mail-order catalogs or locally. If not, there are today such a wide range of gorgeous scarlet to red lilies, your choice is as wide as the modern lily world.

Pick The Lilies You Like Best

I have elected to plant the white Madonna lilies as well as a selection of taller Oriental lilies for several reasons. One is that I believe the original passage must have meant a reddish-colored flower, an anemone or a lily. Also, I like bright red in my garden and Oriental lilies offer bright red color. A third reason is that the other lilies thrive, multiply, and provide an abundance of blooms and dramatic colorful accent areas within a few years. One problem is that you may need to dig up and divide the clumps that form to prevent overgrowth of some lilies. That's a fine way to give plants to friends and encourage them to grow Biblical flowers in their gardens too.

Other Biblical gardeners I have met have adopted the common day lily as representative of the Biblical lilies. They point out that the flower formation is basically identical for all lilies, but day lilies offer easy care and have such a worthwhile multiplier effect that they deserve consideration and use in a Biblical garden. Although each bloom lasts only a day, many colorful new types have been introduced to provide wonderful massed color effects. I must agree with that logic and indeed have included displays at the entrance to our home driveway. They brighten our life every July and well into August.

Lily is the common name for a family comprising more than 250 genera and 3,000-plus species of mostly herbaceous flowering plants. It includes lilies, daffodils, hyacinths, day lilies, and tulips too. Many lilies are adapted to arid climates, as their ancestors were in the Holy Land. During dry periods, food and water are stored in bulbs or in corms, rhizomes, and tubers of other typical Biblical land plants.

In addition to production of seeds and bulbs underground, lilies also sometimes produce bulbils, tiny dark growths at leaf junctions, which apparently are clones of the plant. We have had them form on occasion and others report regular production of these bulbils. They drop off, form tiny roots, and pull themselves into the ground, where they produce new identical plants.

Whichever lilies you decide are appropriate for your garden, dig in well to plant them for years of blooming pleasure. For large white lilies, Dutch Gardens has several dramatic choices. Oriental hybrid lily Casa Blanca grows 40 inches tall. Spaced 12 inches apart, it bears immense, pure white flowers in July and August. Wayside Gardens in Hodges, South Carolina, has Casa Blanca, plus other dramatic lilies that are hardy for gardens nationwide.

Two other lilies, Cassandra and Arena, have white blooms with yellowish tones. Snow Queen, *Lilium longiflorum*, has four to six snow white trumpets in summer on 3-foot-tall plants. *Lily formosanim* is a giant, late-flowering variety with 8- to 10-inch-long fragrant blossoms on 6-foot-high plants This dramatic performer has been shown to be disease-free, prefers sun to light shade in acidic, well-drained soil. It grows in Zones 5 to 9. If in doubt, consult the Zone Map on page 146 or check with local experts to determine your zone. A hardy version of the Madonna lily is available from White Flower Farm, a mail-order firm.

Growing Tips: Plant bulbs in a slightly shaded, moist but well-drained part of your land. Try to give them shady roots, but sunlight around their heads or blooms. One way to achieve this is to use several inches of mulch over soil where you planted bulbs. Another option is to use a ground cover of vinca or ivy to shade the ground. Because the Easter lily has become such a traditional Holy Day plant, you'll find many more details about it in Chapter 8, "Flowers for Holy Days and Holidays." That includes recycling your gift plant for years of blooming beauty in your garden.

ATTRACTIVE ANEMONES—
ANEMONE CORONARIA

"And why take ye thought for raiment? Consider the lilies of the field, how they grow; they toil not, neither do they spin: And yet I say unto you, That Solomon in all his glory was not arrayed like one of these."
—Matthew 6:28–29

"Consider the lilies how they grow: they toil not, they
spin not; and yet I say unto you, that Solomon in all his
glory was not arrayed like one of these. If then God so
clothe the grass, which is today in the field, and tomorrow is cast into the oven; how much more will he clothe
you, O ye of little faith?"—Luke 12:27–28

In early spring in the Holy Land, many thousands of crown
anemones appear. They are dramatic in scarlet but also appear
at times in purple, pink, blue, and white blooms. The fields,
wastelands, and hills throughout Mediterranean areas are colorfully alive with this delightful flower. Many scholars believe
anemones are the flower meant in these passages, not the lily.
After detailed research, I must agree. Let's look at the lily question more closely. It is possible that the lily mentioned in these
Scriptures is a White Madonna lily, since lilies are bulbous
plants and are found in the land of the Bible. Lilies also have a
longtime association with holy days. Many paintings and carvings of the Resurrection and Ascension show the Madonna lily
prominently. In fact, the major artists of the ages, Titian and
Botticelli among them, have all painted the Virgin Mary with a
white Madonna lily. This, however, probably has more to do
with the papal edict that was issued in 1618 than anything else.
This laid down stringent rules concerning the appropriate treatment of certain sacred subjects in art. In this edict the necessity
of showing the white lily in paintings of the Annunciation was
emphasized.

However, the Madonna lily is not especially common in
Israel. In fact, the experts at Neot Kedumim, the world-famous
625-acre Biblical Landscape Reserve in Israel, emphasize that
the Madonna lily, *Lilium candidum*, grows only in the Carmel
mountains, not in the lower fields and meadows. Most authorities now agree that the Madonna lily cannot be the "lilies of the
field" of Matthew 6:28 and Luke 12:27. Another fact must be understood. In various translations of the Scriptures from ancient
languages, the same Greek or Hebrew word has been interpreted

by different translators to mean the tulip, the anemone, or the iris, depending on who did the translation.

Dr. James Moffatt and Dr. Edgar Goodspeed were two astute, dedicated, and well-educated men who undertook translations of the Bible during their lifetimes. What makes their translations different is that they had botanical and horticultural knowledge and therefore dug more deeply into botanical and scientific terms and plant descriptions. With that in mind, it is understandable that they might suggest other plants than early translators did.

Biblical authorities have pointed out that Jesus always chose the most familiar objects from the daily life of the people to illustrate His teachings, as all who have read the Bible realize. This fact leads many scholars to say that the abundant anemone was truly the lily. The anemone did and does grow wild in profusion throughout the land. The anemone, in fact, can be found throughout the Holy Land in many forms today. It blooms from the shores of the Lake of Galilee to the plains and foothills beyond. With its brilliant colors—anemones bloom in scarlet, purple, rose, and yellow—the flower is understandably one of the most conspicuous plants in the Holy Land. Since most authorities today regard the Palestine anemone, *Anemone coronaria*, as the famous lily of the field, we are inclined to accept the majority viewpoint.

Looking at them in terms of botany, anemones are a member of the Ranunculacae family, which has 35 genera and about 2,000 species. Sometimes called windflowers, they seem to prefer windward places, but will grow practically anywhere. One widely cultivated species, the poppy anemone, is a favorite in gardens. There can be five to twelve petals but usually about six in scarlet with a white patch near the flower base. The flowers have the light-sensitive habit of opening in the morning and closing at night. The plants themselves are actually perennial herbs with an underground rootstock and have radical, deeply cut leaves with elongated flower stems.

Today, we can treat ourselves and friends to even greater glories of anemones—more even than the people of Biblical times could. If you prefer bright red anemones, select De Caen Hollandia, a special variety now widely available. Plant them in groups. Each tuber produces four to six single flowers at intervals of 1 to 2 weeks. For bright blue anemones, select De Caen blue poppy, which adds rare blue blooms to your summer garden.

You also have another option, the Japanese anemone, *A. japonica*, which is an autumn-blooming species. Adding these to your garden lets you enjoy both a spring and fall display of anemones. Plants grow 12 to 15 inches tall. Space corms 2 to 3 inches apart. They grow nicely in horticultural Zones 4 to 10 but are outdoor hardy in Zones 6 to 10. You'll find a map of the United States horticutural zones on page 146 of this book. Plant these anemones in groups in borders or beds. They also make excellent displays with other flowers, and you can even use them to brighten vegetable gardens.

Growing Tips: Anemones grow well from tiny tubers, which should be soaked for 24 hours in warm water before you plant them in the ground, after all danger of frost has well passed in the area in which you live. Anemones are useful as ground covers and prefer light shade. Use them in rock gardens, under shrubs, and beneath trees. Plant tubers about 2 inches deep and 2 to 4 inches apart in cool, moist, well-drained soil. In Northern states, plant anemones in the early spring or fall. In Southern areas, plant in the fall and cover with mulch during the first winter.

Anemones sprout readily with adequate moisture, but will do well in somewhat dry soils. Although they thrive wild in rocky, poor soil, for best results in gardens they should have compost and fertilizer periodically. Cover with mulch to save moisture and thwart weeds.

Like narcissus, anemones will multiply every year. If you wish to add them to your garden, there are many types available today, including double-flowering with more profuse blooms. However, the poppy-flowered, single anemone should be your first as most representative of a true Biblical flower. Be sure

when you shop, locally or from catalogs, that you order anemones certified for hardiness in Zones 3 to 5. Most are this
hardy, but it is best to check.

CROCUS—*CROCUS SATIVUS*

"Spikenard and saffron; calamus and cinnamon, with
all trees of frankincense: myrrh and aloes, with all the
chief spices . . ."—Song of Solomon 4:14

"Your shoots are an orchard of pomegranates with all
choicest fruits, henna with nard, nard and saffron, calamus and cinnamon, with all trees of frankincense,
myrrh and aloes, with all chief spices."—Song of Solomon 4:13–24

"The desert shall rejoice and blossom; like the crocus it
shall blossom abundantly, and rejoice with joy and
singing."—Isaiah 35:1–2

Fortunately, in tracing the roots of Biblical plants, it is occasionally possible to find one identification on which almost all
authorities—Biblical scholars and botanists alike—agree. Although
saffron—in Hebrew, *karkom*—is mentioned only once in the
Scriptures along with spikenard and cinnamon, there is no
question about what plant is being discussed. Saffron comes
from one particular crocus, the saffron crocus, so we can be as-
sured that the crocus is an authentic plant of the Scriptures.

Saffron Is Historic Dye

The saffron crocus, *Crocus sativus*, is native to Asia Minor and
Greece, and is also found in other Mediterranean countries.
After the stigmas and styles, those distinctive reproductive parts
of the flower, are gathered, they are dried in the sun, then
pounded into small cakes. This expensive product is used primarily as a yellow dye, and also for coloring in curries and
Oriental foods. Saffron was a valued dye in the time of the Bible,
as it is today, especially in India and other Oriental countries far

beyond the borders of the Holy Land. Talmudic sources focused on karkom depict a plant whose flowers, the stigmas, were collected for giving food and fabrics the typical saffron yellow color. Considering the fact that it requires at least four thousand stigmas and upper portions of the style of the blue-flowered Saffron crocus to make an ounce of the yellow dye, it is one of the most expensive coloring products. No wonder it was so valued in the time of the Bible!

Crocus, a genus that comprises closely related plants with a single name, are hardy perennial plants and a member of the iris family, Iridaceae. Crocuses produce a single tubular flower from a corm. They have grasslike leaves that grow in a rosette pattern. The common fall crocus, *C. sativus*, has bright lilac-blue flowers. Spring-flowering crocuses, including the early-flowering, so-called Dutch crocus, are available in violet to purple hues and in various other colors.

Crocuses Grow Well in the Holy Land

Actually, many types of crocus grow in the Holy Land, as they did before the birth of Christ. You can find white and pink, blue and yellow, as well as lilac and purple crocuses in the fields, on rocky hillsides, and along roadsides. However, only the blue-flowered ones produce saffron. These must have grown in great abundance to have withstood the constant harvesting that the demand for saffron created. In ancient times, old records reveal that saffron was scattered during wedding ceremonies and mixed with wine. It has been used to make perfume and to color confections as well. Some old texts indicate that saffron was given medicinally as a stimulant and with spices and flower petals as scents to perfume rooms.

In other passages of the Scriptures in the translation of Dr. James Moffatt between 1922 and 1925, he takes pains to change some of the wording of the Song of Solomon. Where, in 6:4 it is written in the authorized Version, "Thou art beautiful, O my love, as Tirzah, comely as Jerusalem . . . ," Dr. Moffatt is of the view that no love song would compare a young maid to cities.

Therefore, he renders the passage as "You are fair as a crocus, my dear, lovely as a lily of the valley." While it may seem more appropriate to compare a young girl to flowers rather than to cities, other scholars believe that Dr. Moffatt has taken some liberties in his translation.

However, if you feel as Dr. Moffatt did, that a crocus is a more fitting comparison, you might think about what other Biblical scholars have said. Many believe that if the Song of Solomon was actually composed by Solomon, it was done somewhere in the hills and mountains of Lebanon. There, botanists report, four different species of crocus do occur in profusion. Whichever way you wish to vote, you can enjoy the beauty of modern crocuses around your home and garden as the ancients did.

The saffron crocus closely resembles our springtime crocus, but actually it blooms in its native habitat in late autumn. The saffron crocus is a tiny plant with a subterranean corm. It produces several narrow leaves and one or more large, bluish-lilac flowers. You may choose for your garden the purple-blue spring crocus or the purple fall crocus, which also thrives across northern regions of America.

Crocuses Are Easy to Grow

Crocus bulbs are easy to plant. No garden should be without these cheerful welcomers of spring. By planting the largest, earliest-flowering varieties, you often will be greeted by their colorful blooms even before all snow has melted. Today's crocus varieties have been specially developed for more profuse blooms and showier colors than the original wild species. Truest in appearance to the Scriptural plants are the blue ones, whether you plant for spring or fall bloom. The fall types multiply rapidly without care or trouble.

Growing Tips: Plant autumn crocuses in the spring, in sun or light shade areas. As a perennial, they become a permanent part of your plantscape beneath trees, under shrubs, in beds, and along pathways. Once planted, they increase year after year to produce profusions of blooms. Spring-flowering crocuses should be planted between mid-September and early November, before the ground freezes. New varieties produced by plant hybridizers and offered by mail-order firms are related to the crocus of the Scripture and offer larger, more abundant blooms.

Plant bulbs 3 to 4 inches deep in clusters or groups. Once planted, they are so persistent that even an amateur gardener can obtain perfect results. Most crocuses multiply year after year by forming bulblets from their corms during the growing season. For best results, it pays to obtain larger corms, which cost a little more but produce bigger blooms and get your garden started better.

CYCLAMEN—*CYCLAMEN PERSICUM*

"Consider the lilies how they grow: they toil not, they spin not; and yet I say unto you, that Solomon in all his glory was not arrayed like one of these."—Luke 12:27

Most likely, the cyclamen is another prime contender as a "lily of the fields" according to botanists. When earliest translations of the Bible were made, there was no botanical nomenclature. The term "lilies" was loosely translated to mean a variety of common flowers. Two forms of cyclamen were common in the Holy Land in ancient times. They are believed to be among the flowers mentioned in Luke 12:27. One variety, *Cyclamen persicum*, is also called by a common name, alpine violet. It grows wild in stone walls, among rocks, and crevices, and along roadways in the Holy Land today. This plant prefers an acidic soil and some shade from the afternoon sun. Cyclamens offer a long blooming period, often of several more weeks compared to most other flowers. Now that hardy varieties are more readily available, they are prime candidates for every Biblical garden.

Blooms begin in November in their native habitat in the Holy Land and may last a month or more. They have been known to continue flowering in the Upper Galilee area into early May. Cyclamens elsewhere in the Mediterranean zone, in which they are well adapted, bloom through the winter with vivid pink to white blossoms. These blooms are often likened to miniature butterflies poised on slender stems of the plants. Don't confuse these outdoor cyclamens with the florists' version. The florist type are good indoors but won't survive outdoors.

During the Christmas season millions of people now purchase cyclamen plants for themselves and to give as gifts. Cyclamens have become popular throughout the year too. Popularly known as Solomon's crown, it recalls the Jewish king who built the first temple in Jerusalem. The potted plants from florists closely resemble the wild ones in the Holy Land today. Unfortunately, many who receive these as gifts are unaware of their Scriptural roots or native habitat in the Holy Land. With this florist-type indoor cyclamen, you can extend the blooming season for many weeks in your home. We have a florists' cyclamen that has bloomed periodically and for ten to fifteen weeks at a time for several years. Key tip is to move it out of the sun's rays in the afternoon and back to its window each morning. Too much sun wilts it badly.

If you wish to enjoy a tender florist's type cyclamen outdoors for a summer, follow these tips. Place the pot outside during June in a cool, shady location. Midafternoon shade is important to help provide lower temperature. Maintain a moist soil condition within the pot but be sure that excess moisture can drain away. During the growing season, fertilize the cyclamen every two weeks with a liquid fertilizer solution. Follow label directions for the brand you use to avoid overfeeding. If you place the pot in the ground, turn it every two weeks. This will break off any roots that may grow through the drainage hole. This is important to prevent shock to the plant when it is moved indoors at the end of summer. If you prefer, you can double-pot your cyclamen in a clay pot surrounded by peat or sphagnum moss inside an outer decorative container. That works well for porch, balcony, or patio display.

By mid-September, move the plant indoors to prevent any chance of frost damage. Keep it evenly moist and fertilize it every third week. Too heavy fertilizing could produce leggy plants with reduced flowering. By late fall, more buds should form on the flower stalk. With proper watering and feeding during the summer outdoors, you may get some blooms then and the plant will be healthy and ready to produce abundant blooms for the Christmas season.

For permanent display in outdoor gardens, *Cyclamen persicum* is available. It closely resembles the cyclamen of the Holy Land. These are cultivated in Israel and imported by Van Bourgondien, a long-time mail-order firm specializing in bulbous plants. This hardy cyclamen needs a soil with lots of sand and composted leaf humus. Keep in mind that even so-called hardy cyclamens are recommended for Zones 6 to 9. You may need to do some experimenting and risk-taking in Zones 3 to 6 as you work on cyclamen additions to your gardens in colder areas. Some seem hardier and make it through a mild winter or two. Mulching is important in the fall to insulate the ground from freeze and thaw cycles that can damage even hardy types.

Cyclamen, a genus of decorative plants, is in the primrose family, Primulaceae, and native to Europe and Asia. It is becoming more popular for gardens and as house plants in the United States. Cyclamens are perennials and rise from corm-like tubers. The plants have butterfly-like flowers on stalks that are 6 to 8 inches high. Petals are bent sharply downward and leaves are marked with white along the veins.

The most common cyclamens in American gardens are hybrids of the Persian cyclamen with flowers in various reddish shades on stalks 6 to 8 inches tall. The deep red variety is classified as *C. europaeum*. The one with pink and white is usually classed as *C. neapolitanum*. Fortunately, a cultivated version of the true Holy Land cyclamen, *C. persicum*, is now being imported and is available to all.

Growing Tips: The hardy cyclamen, *C. persicum* is available from Van Bourgondien on Long Island, New York, at 1-800-622-9997 and is from cultivated stock in Israel. These corms are grown outdoors under shade cloth in pots. Plant corms concave side up ½- to 1-inch deep and 6 to 8 inches apart in moist, well-drained, and slightly acid soil in a shady location. Moisten soil regularly for best results.

Another hardy cyclamen, *C. neapolitanum*, is a gorgeous ground cover that begins flowering in late summer and early fall. Good corms are 5 inches in circumference. Space them 8 to 12 inches apart. After flowering, leaves continue to develop and may stay green until next spring before going dormant in summer, according to the Dutch Gardens specialists, another leading flower bulb importer. The *C. neapolitanum* is hardy and ideal for gardens and naturalizing. It is good for rock gardens or shaded areas under trees and shrubs. *Cyclamen cilicium*, another hardy cyclamen, performs well as a ground cover.

HYACINTHS—*HYACINTHUS ORIENTALIS*

"My beloved is gone down into his garden, to the beds of spices, to feed in the gardens, and to gather lilies. I am my beloved's, and my beloved is mine: he feedeth among the lilies. Thou art beautiful, O my love, as Tirzah, comely as Jerusalem . . ."—Song of Solomon 6:2–4

Hyacinths are one of the most welcome delights among Biblical plants because of their aroma. The flower originated in Palestine and the Middle East, and ancient writings reveal that they were cultivated by some of the world's earliest gardeners. However, their true significance and identity as a flower mentioned in the Scriptures has been debated for centuries. Hyacinth is the common name for a genus of plants of the lily family, Liliaceae. They have been cherished as garden and house plants for centuries in temperate and tropical climates around the world. Hyacinths are bulbous plants with sword-shaped leaves. Flowers appear on long spikes and have three petal-like sepals, three petals, and six stamens.

In the Song of Solomon we find the most quoted Scriptural mention of lilies in the Bible. Some scholars believe that this reference to "lilies" actually means the more numerous anemones that grow profusely in the Holy Land. Other authorities believe that the proper botanical interpretation of the lilies mentioned in this passage from the Song of Solomon actually refers to hyacinths, *Hyacinthus orientalis*. The debate has continued through the past millennia and persists today.

Perhaps a review of this Scriptural passage, as countless others have done through the centuries, will be helpful to an understanding of this Biblical flower's rightful place in Biblical gardens. The passage reads: "My beloved is gone down into his garden, to the beds of spices, to feed in the gardens, and to gather lilies. I am my beloved's, and my beloved is mine: he feedeth among the lilies. Thou art beautiful, O my love, as Tirzah, comely as Jerusalem . . ." This is the translation in the King James Version.

After pondering this passage, scholars and botanical have agreed that it is not possible for the plant to be a lily because it is not a plant of the lowlands. Other authorities also seem to be most confused trying to equate the lilies of the fields with lily of the valley. That cannot be true because the lily-of-the-valley plant cannot be traced back botanically anywhere near the land of the Bible.

The Goodspeed translation actually makes a point of focusing on the hyacinth in the Song of Solomon 2:1–2. "I am a saffron of the plain, a hyacinth of the valleys." As a meticulous scholar and dedicated researcher, Dr. Goodspeed undoubtedly had done sufficient scientific homework. His identification of the lilies of the valley as hyacinths seems logical and plausible. He goes further in his translation of the Song of Solomon 6:3, declaring that: "I belong to my beloved and my beloved to me, who pastures his flock among the hyacinths."

Hyacinths Are Prolific in the Holy Land

What makes many scholars—Biblical, botanical, and horticultural alike—believe that the hyacinth, *Hyacinthus orientalis*, is the true plant is its wide natural range in Palestine and other areas of the Middle East. It flowers profusely in the wild, bearing deep blue, highly fragrant spikes. During the blooming time in Israel, the hillsides in Galilee are blanketed in blue from hyacinth blooms. Through the generations, this plant has been one of the most easily recognized wildflowers and its heady perfume has made it a favorite throughout Europe and much of the world. Hybridizers, dedicated plant-breeding experts, have created fuller blossoms and preserved that distinctive fragrance. It is a "must grow" plant for gardens, especially upwind from your outdoor entertaining areas.

Aeons ago the fragrance of hyacinths in ancient Greece and Rome inspired poets and gardeners alike. Today, you have a wide color choice from deep blue and purple to pinks, reds, and whites. They respond well every spring when grown in fertile garden soil in beds and borders. As massed plantings they perfume the air with intoxicating aromas every spring.

Cultivated hyacinths, commonly called the Dutch hyacinth, are originally native to Greece and Asia Minor as well as the Holy Land. Modern horticultural varieties have been hybridized so they are more profuse with larger blooms, and greater fragrance. Because they have such profuse blooms, modern Dutch hybrid varieties don't realistically resemble their wild ancestors, which still bloom in the Middle East but have simple blossoms.

However, Roman hyacinths, *H. orientalis* variety albulus, produce loosely flowered, graceful spikes from each bulb. Frankly, I think these more closely resemble the wild native hyacinths of the Holy Land. These types are mainly grown indoors, or outside only in very mild climate areas. These so-called Roman hyacinths are produced in Italy and southern France. Dutch hyacinths, the hardy perennials for outdoor use, are produced in the Netherlands. Both types flower in spring.

You can obtain good hyacinth bulbs in local garden centers or departments that will provide excellent flowers on sturdy stems. For largest blooms, you should order mammoth bulbs 18 to 20 cm in size which produce glorious displays. Like tulips and daffodils, hyacinths have been hybridized to exceptional quality and growth performance by plant breeders. Mail-order garden firms, especially those that specialize in bulbous plants from the Netherlands, the world's premier commercial producing area for bulbs, offer a wide selection. You'll find a listing of them in

Chapters 11 and 12. Several leading firms offer colorful, free catalogs featuring all types of bulbous plants suitable for Biblical
gardens.

Dig Deep to Grow Hyacinths Well

Hyacinths will produce blooms in any type of reasonably fertile
soil that is well drained. They, like most other bulbous plants, do
not tolerate soggy root conditions. Because they are spring-
blooming, hyacinths should be planted in fall, between the middle of September and late October, before the ground hardens in
northern climates. In southerly areas you may plant hyacinths
into mid-November.

For hyacinths you should dig more deeply than other bulbous plants require. They prefer 12 inches of well-drained soil
under the bulb since roots grow about 10 to 12 inches deeper
down in the soil. It pays to dig down deeply, plant well, and
enjoy this most fragrant of Biblical plants for decades to come.
Plant actual bulbs six inches deep and give them sufficient room
by spacing 5 to 8 inches apart in beds. Hyacinths can remain in
that location for years.

If you want really massed color, plant bulbs closer to one another but be prepared to compensate for close planting with
extra fertilizer each year. In outdoor gardens, be sure to let foliage ripen after plants have bloomed. The leaves must manufacture food and store it in the bulbs to produce and nourish next
year's plants. A light mulch helps prevent bulb heaving in cold
winter areas.

Try Water Glass Hyacinth Fun

Hyacinths have another marvelous advantage. They produce
magnificent blooms in special hyacinth-growing glasses. You
can obtain kits of these giant bulbs that nestle in the tops of
these glasses at some of the major chain stores like Wal-Mart and
Home Depot. Local garden centers also often carry them. This
makes a blooming family fun project from late October to De-

cember to give you the color and fragrance of a Biblical garden indoors.

Before setting the bulb in the hyacinth glass, clean the root base of all old residue and roots. Fill the glass with clear, cool water until the root base touches the water. Then place the glass in a cool, dark closet or cellar until the top growth is 5 inches above the bulb. This will usually occur in 6 to 8 weeks. Keep plants where temperatures don't exceed 50 degrees F or fall below 40 degrees F.

When the bulb has sprouted well, move the plant to a north window in your home or apartment. Add warm water if necessary so it covers the roots that have formed in the lower portion of the container. When you place this started hyacinth in your living room, keep it away from direct sunlight. The flower spike should grow tall and sturdy and reward you with its giant bloom and the fragrant scent for a week or more.

You can also plant and force hyacinths in pots or tubs by themselves, as groups, or combined with other Biblical flowers such as daffodils, crocuses, and tulips. You'll find details in Chapter 6 about forcing bulbs, i.e., tricking or coaxing them to bloom before their normal time, for yourself or as gifts for friends.

Wild hyacinths, *Hyacinthus orientalis,* bloom in blue, purplish, and bluish white. They can be seen blossoming each year in the Holy Land and are a sure plant of the Scriptures, as you will find in Solomon 6:2. Wood hyacinths, *Scilla campanulata,* are a type that came from Spain originally, but actually look much like the wild ones. These late spring blooming bulbs make an attractive sight in wooded areas. Plant bulbs 3 inches apart in Zones 3 to 9.

The Dutch have been world masters at creating gracious and dramatic new varieties of flowers from bulbous plants. They gave the world a plethora of daffodils and tulips. Few realize how much progress the Netherlands plant hybridizers have made with hyacinths. There are many brilliant colors today from the blues and reds to hot pink and orange.

Hyacinths thwart squirrels, chipmunks, and even deer. There

are irritating chemicals contained in the bulbs, which even give some people an itchy rash. If you have sensitive skin, wash well after planting them. But be happy that the chemical deters deer and squirrels, too.

IRIS—BLUE FLAG—*IRIS VERSICOLOR, IRIS ASTROPUPUREA*

"I will be as the dew until Israel: he shall grow as the lily and cast forth his roots as Lebanon."—Hosea 14:5

Is the iris really mentioned in Scriptures as a Biblical plant? That question has been asked for ages, it seems. The debate focuses on this single passage from Hosea 14:5. In this passage that mentions lily, many scholars believe it probably identifies yellow flag, or yellow water iris, commonly called the yellow flag iris today. Many species of iris grow on hills and mountainsides, in fields and even drier areas. However, it is the yellow flag that displays its beauty at the margins of streams and waterways, often in masses of color. Because it flourishes along the edge of streams where you also find poplar and willow trees growing, botanists believe the yellow flag may be the plant indicated in this passage from Hosea. The Eurasian yellow flag, *I. pseudacorus*, actually is one of the most common wild irises in the world, blooming in June and July in temperate climates.

In the United States, both the wild yellow and blue flags also favor such environments. In fact, blue flag irises also grow wild in the Holy Land, also in moist soil areas. Consulting with my best advisors, I've collected more facts and tips to help you add both these irises, blue and yellow, or close look-alikes among hybrid types, to your garden. Iris is the common name of a family Iridaceae of herbaceous flowering plants. More than 1800 species exist in more than 90 genera. These handsome flowers grow from underground stems, which are called rhizomes. Since they have food storage ability in these roots, growing irises is relatively easy. For accurate Biblical garden focus, we probably should concentrate first on the iris that most closely resembles iris in ancient Palestine, which thrive in wet areas of Israel today, the blue flag iris. Other cultivated hybrids offer us options.

Iris is a symbol of undefeatable believers, an encouragement to those of us who face a share of testing in life. This plant survives great dryness and yet it thrives immersed in water as one of the "lilies who take root like the roses of Lebanon," according to another translation of Hosea 14:5. That makes them a good choice for gardens everywhere.

Iris blooms have been admired throughout recorded history. The fleur-de-lis pattern has been captured in tapestries, carvings, and ceremonial banners from ancient times through the present. The iris blooms are depicted in many designs and decorations today, from jewelry to clothing and on silverware patterns.

The large blue flag iris shoots its lance-shaped leaves from wet meadows and pastures in early spring. Its neatly furled buds burst into blue blossoms on 16- to 30-inch stems. Although this iris prefers marshes, wet pastures, and banks or streams or ponds, it will flourish is moist garden soils or boggy areas if provided regularly with ample moisture.

Flowers are composed of a floral envelope, a perianth with six petal-like segments, three or six stamens. The inner three segments of the flower, called standards, are erect and narrowed at the base. The other three are narrowed and usually droop and are called falls. The beard in special bearded irises consists of colored hairs on the upper surface of the falls.

Among iris, one type has creeping underground stems or rhizomes. That includes the bearded or German irises and the Japanese and Siberian or beardless iris. The beard is that distinctive addition to the iris bloom in German irises. The second group has bulbs that horticulturists describe as modified underground buds and includes the Dutch, Spanish, and English varieties. Catalogs specify the height of the different types of irises available today.

Among Siberian irises, Caesar's Brother is one of the tallest and most elegant purple irises and is essential free of pests and disease. It matures 3 to 4 feet tall and does well in Zones 4 to 10. It blooms in late spring to early summer. Silver Edge is another good blue-colored iris. Your choice in various mail-order catalogs is much wider, from Iris SkyWings or Iris Steve from Van Bourgondien to Wayside's Swank and Dreaming Spires. Cultivated irises tend to spread, extending their blooming scope. Wild blue flags also reward you more profusely each year as they spread, sending up their violet to slate blue blooms.

Growing Tips: There are different types growing wild in America, but all wild ones like their domesticated and hybrid relatives can be transplanted easily by moving entire clumps or dividing the rhizomes with a sharp knife. Naturally, always ask the property owner's permission before collecting any plants from the wild. Also keep rhizomes moist during travel. Then, replant rhizomes at the soil surface, just tucked in enough to stay in place. Pick a moist area that matches their original habitat.

Iris rhizomes must be planted right at the surface, never beneath soil. If planted deeply, iris won't thrive. That's true with cultivated as well as wild types.

Cultivated look-alike Siberian and most other types of irises need well-drained soil. They'll grow in sandy to clay and prefer full sun locations. July to September is best planting time. Space rhizomes 1 to 2 feet apart. Set with tops just beneath soil surface, spreading roots well. Growth proceeds from the leafy end of the rhizome, so place that in the direction you want plant growth. Water well and periodically until plants get a firm roothold. For dramatic show, plant several or a group of rhizomes together. Add liquid fertilizer before and after bloom for sturdier plants, better blooms. As clumps become crowded, simply cut rhizomes in July or August and replant elsewhere or trade with other Biblical gardeners.

IRIS YELLOW FLAG—*IRIS PSEUDACORUS* AND LOOK–ALIKE OPTIONS

"I will be as the dew until Israel: he shall grow as the lily
and cast forth his roots as Lebanon."—Hosea 14:5

As we read the Bible, especially in Exodus 30:23, "Take the finest spices: of liquid myrrh five hundred shekels, and of sweet-smelling cinnamon half as much, that is two hundred fifty, and two hundred fifty of aromatic cane," from the New Revised Standard Version, we may be tempted to identify the aromatic plant as some type of sugar cane. That was not known in Biblical times.

However, caravans carried a fragrant cane-like root, a rhizome in the spice trade that we now believe actually was the sweet flag, *Acorus calamus*. Originally it was from Central Asia, but has become wild and nearly native in the Holy Land and other temperate parts of the world. Its name, sweet flag, tells us that technically it is not an iris, although it closely resembles one. The caravan trader's root probably is a waterside plant that belongs in the Araceae family. It has smaller yellowish flowers and fragrant sword-shaped leaves.

For practical purposes, it is easier to find and grow the look-alike, the yellow flag iris. It is easily grown in mud or wet soil along lakes, pools, or streams. Plant rhizomes in early summer. They will usually take root quickly and spread to give delightful displays. If they overgrow and crowd out other plants, simply dig out the surplus and give it to other gardeners. You can control them by growing them in sunken pots or tubs of rich composted soil that are set into streams or ponds, just as you can do with water lilies.

Iris pseudacorus variegata is available from Van Bourgondien, one of America's leading mail-order firms specializing in bulbous plants. It is located on Long Island, New York. Other yellow iris are appropriate look-alikes and I endorse them for ease of growth and longevity too. Butter and Sugar Siberian irises are hardy perennials featured by several mail-order firms including Wayside Gardens. They retain the simplicity of the original

Biblical iris because they, too, are not bearded. These Siberian irises enjoy a sunny location and normal soil conditions. They also thrive under moist conditions so a pond or stream side is a good option. These plants are essentially free of pests and diseases and will grow for many years without needing to be divided. Another option is Snow Queen, also available from Van Bourgondien and Dutch Gardens.

Iris pseudacorus was originally from Asia. It is a decorative and hardy-growing plant that forms large clumps of flowers with yellow iris blooms and matures 3 to 5 feet tall. It is a worthy substitute for sweet flag, or grown with it in moist, watery areas.

Growing Tips: Plant in muddy shores of streams or ditches, or keep soil moist for this special plant to perform. Divide rhizomes with a sharp knife and plant at the soil surface. Move entire clumps if available and set into wet soil areas. See additional planting directions in the preceding blue flag iris section. Those culture tips can guide you toward growing Siberian and other types of look-alikes in your garden, especially if you have no water or wet areas.

No doubt, yellow flag and blue flag iris brighten a water garden that can also include other Biblical plants such as rushes and reeds. However, without wet areas, the look-alikes are an appropriate replacement to represent this Scriptural flower. Check through mail-order catalogs to find the best-looking yellow iris as a representative Biblical iris.

NARCISSUS—*NARCISSUS THIAZIDE* (DAFFODILS)

"The wilderness and the solitary place shall be glad for them; and the desert shall rejoice, and blossom as the rose."—Isaiah 35:1

"I am the rose of Sharon, and the lily of the valleys. As the lily among thorns, so is my love among the daughters."—Song of Solomon 2:1–2

As you will have noted from earlier comments, debate about which Scriptural passage and word actually mean a specific flower has been ongoing among Biblical and botanical scholars for centuries. The narcissus is thought by some to be the Biblical lily. I, too, believe that the narcissus fittingly represents the flowers of at least one of the above passages.

The narcissus is a member of the Amaryllis family, which includes 65 genera and 800 species. For reference, the yellow daffodil is classed as *N. pseudonarcissus,* and has a deep, trumpet-shaped crown; the polyanthus as *N. thiazide,* and jonquil as *N. jonquilla.* Jonquils typically have yellow flowers with short crowns. Poet's narcissus has pure white petals and a shallow, wrinkled, reddish crown. Polyanthus, closest to the Holy Land wild type, has small white or yellow flowers and short coronas.

Two species of narcissus are native to Israel. *Narcissus thiazide* often is one of the most striking early-blooming flowers seen there. The wild *narcissus thiazide* grows in the damp soil of the plains as well as among shrubs and rocks in the hills and even into the Negev Desert area. It flowers in November in its native habitat and goes dormant in the dry season.

The wild natural narcissus has flowers which open successively and spread horizontally. They consist of the leaves, the tall stem, and the spreading white six-petal lobes with erect lemon yellow crown. It is duplicated by many hybridized types of narcissi for our gardens. Many will take root and grow as natural clumps in lawns, under shrubs and trees, and give landscaping a so-called naturalized look. We like to use them in wilder, naturalized settings on our land. Under trees, on banks, in clusters in rock gardens, and on slopes, a variety of different narcissus offer us exciting opportunities for displays. All trace their roots to that remarkable, hardy Holy Land flower. Fortunately today, *Narcissus thiazide,* is available from Van Bourgondien. It offers several types that closely resemble the wild flowers seen each year in the Holy Land, as the exceptional photos in the insert show in glorious color.

Narcissi were notable in ancient times, mentioned in the Scriptures, and still grace the wild areas of the Holy Land today each spring. The plants have long, slender leaves growing upward from the bulbous root. Flowers, borne single, usually have six petal-like sepals, a corolla of six united petals, six stamens, and a solitary pistil. The cup-shaped crown, called the corona, rises from the inner surface of the corolla. Flowers are usually yellow, with combinations of shades.

Hybridizers who create new varieties of plants have made their talented contributions to this plant too, creating hundreds if not thousands of variations today. You can put them to dramatic accent use in your garden, naturalize them in meadows, plant them under trees or shrubs, or create giant flowering crosses in green lawns for spring display. Narcissi are also called daffodils by many and are one of the first harbingers of spring to bloom right after crocuses do. Their large flowers appear early and they thrive in moist soil, doing best in temperate climate areas.

You can obtain close look-alikes to *N. thiazide* from mail-order firms, garden centers, and local florists in spring. Paper white narcissus and a similar one called Chinese sacred lily can both be grown outdoors, but do even better as container-grown, forced plants. Since both are close in appearance to the wild narcissus of the Holy Land, you may wish to try these first. Growing these for Sunday school projects also is a thrilling, eye-opening, and learning experience for youngsters. Some Biblical gardens have outreach programs featuring potted or container plants. The adults see it as a way to teach responsibility and give joy and pride to each growing child, plus an opportunity to impart some of the Scriptures to young gardeners.

Bulbs and planting mix or complete kits can be obtained in many chain stores. Paper white polyanthus bulbs should be planted on top of a gravel and peat moss mix. Six bulbs will fill a dish 6 inches in diameter. Place water in the dish as you would in forcing crocus or hyacinth bulbs. Keep it resupplied to the level of the bottom of the bulb. Within weeks, you'll be rewarded by large, sweet-scented clusters of white flowers with their diminutive light yellow cups. You may also find other types of narcissus for growing in pots. A golden yellow variety, Soleil d'or, is similar but flowers are deep yellow with an elegant orange-yellow cup.

In Chapter 6, you will find more details about growing various Biblical flowers in containers and also coaxing them to bloom when winter winds still wail. For children, growing Biblical flowers in pots and tubs can be one of the most exciting gardening projects ever.

Growing Tips: Dig deeply for fall planting so roots can grow and establish the plant before cold weather arrives to freeze the ground. Plant narcissus bulbs at least 4 to 6 inches deep between September and mid-November. The new bulb planting tools are handy. Simply place the cylindrical tool on the ground and push down to its full length, which is the correct planting depth. Pull up, insert a bulb or two and replace the sod plug. A dibble works well too or you can dig up areas, set bulbs, and cover them with soil. Water well as you plant and moisten again weekly so they begin their life in your garden properly.

Narcissi are beautiful in massed springtime displays. They also are appealing in beds and borders, under trees and shrubs. My favorite use is naturalized plantings with wildflowers or in meadows and lawns for spring color. As with Star of Bethlehem, which will be discussed below, narcissus bulbs contained alkaloids that are poisons and can cause severe vomiting and convulsions. Exercise caution as you plant them so children and pets are kept safe.

STAR OF BETHLEHEM—DOVE'S DUNG— *ORNITHOGALUM UMBELLATUM*

"And there was a great famine in Samaria: and, behold, they besieged it, until an ass's head was sold for fourscore pieces of silver, and the fourth part of a cab of dove's dung for five pieces of silver."—II Kings 6:25

Today, as in ancient times, the star-of-Bethlehem grows wild in the Holy Land, dotting fields, hillsides, and other stony places with its bright white blossoms each year. It may seem odd to consider that a plant was called "dove's dung" in the book of Kings. However, many scholars explain that the phrase identifies how these lovely white flowers appear when profusions of them bloom in their time, as white droppings from doves. People visiting the Holy Land when these white-starred flowers burst forth in profusion with the first rains of the season have agreed that they must indeed be the plant mentioned in the Scriptures.

The star-of-Bethlehem, *Ornithogalum umbellatum,* is a bulbous-rooted herb member of the lily family. It has erect, grass-like, linear leaves with a white band along their midrib. The leaves grow in a clump from a bulb that looks like an onion. Numerous white, lily-like, star-shaped flowers bloom in spring on stalks 10 to 12 inches high. The star-of-Bethlehem has been a favorite in American gardens for years. These delightful early flowers have often escaped to naturalize in grasslands and meadows in the eastern and central United States.

Growing Tips: You can grow star-of-Bethlehem easily from bulbs. Plant them in reasonable fertile soil, 2 inches deep. They do best in natural settings, beneath trees and shrubs that put out their leaves after this flower has had its early spring blooming time. As a spring-flowering plant, bulbs should be planted in the fall, the same time you set out crocus, hyacinths, narcissi, and tulips. Come spring, the free-blooming star-of-Bethlehem will delight you with its showy clusters of sparkling white flowers and white-striped leaves. Be careful when handling because the bulbs contain toxins that can cause problems. Keep away from children and pets.

TULIP—*TULIPA MONTANTA*

"I am the rose of Sharon and the lily of the valleys."
—Song of Solomon 2:1

"The flowers appear on the earth, the time of singing has come and the voice of the turtledove is heard in our land."—The Song of Solomon 2:12

Earlier in this chapter, and elsewhere in this book, you will have read about the debates about which flower is actually intended by various Scriptural passages. As you will have noted earlier, the debates that have been waged for many decades, perhaps even centuries, concerning the true identity of the rose in the Scriptures continue to this day. Some experts still favor the fall crocus while others adhere to their belief that it is a narcissus. There is another group of gardeners who are adamant that the rose in this Scriptural reference is the tulip, *Tulipa montana*. Others agree on the tulip but argue for *Tulipa scharonensis*.

Trying to refute the belief that the "rose" of the Song of Solomon could really have been one of the plants we call a rose today, some botanists point out that there are no true roses in the climate of the Holy Land. That is not accurate, since old-fashioned wild roses are known in the mountains of Lebanon. You will find details about them in the following chapter. In fact, I strongly recommend that Biblical gardeners do consider growing wild roses as part of their representative Holy Land landscape.

More recent research at the Hebrew University in Jerusalem and elsewhere has focused on tracing the history of plants of the land of the Bible and particularly the background of the attractive tulip plant with its scarlet flowers as the rose. Since this species of tulip is indeed commonly found in the mountainous regions of Lebanon and Syria too, it may truly be the rose of the Song. That wild tulip of the Holy Land bears a single, bright red flower. Many blooming together create an appealing sight in the wild where they still grow today.

Ancient records of other civilizations tell us that tulips were cultivated ages ago in Turkey. The name "tulip," some linguists believe, comes from a Persian word meaning "turban." However you look at these plants, there is no doubt that their roots are set deep in antiquity and they still grow wild in modern Israel. Since they do and are such lovely flowers, they deserve a prominent place in this book and this chapter of Scriptural reference flowers.

Other equally learned scholars have their view that the rose in this passage from the Song of Solomon really is intended to identify the mountain tulips, *Tulipa montana*. Although there are many other red flowers that grow in the Holy Land, and grew there in antiquity, the mountain tulip is the prevailing species of the genus Tulipa growing in Israel still today. Their striking beauty led people to cultivate these plants from earliest times.

Tulip is the common name for a genus of spring-flowering, bulbous flowers of the lily family, Liliaceae. About 80 species exist and are native to the Mediterranean region. Thousands of glorious and dramatic hybrid types have been created by plant breeders to grace our gardens. The original mountain tulip is a bulbous herb with several lancelot leaves centered at the base of a 6- to 8-inch-long stem. The large crimson red flower lives for a week or so.

Through the centuries, tulips perhaps have had more attention from plant breeders than almost any other flower, except roses. The tulip craze that hit Holland and the world in the l600s drove prices of special hybrid tulips to fantastic levels. Collectors paid hundreds of dollars for single bulbs of newly developed, decorative strains and varieties. Eventually that tulip craze bubble burst.

Fortunately, prices for tulips today are reasonable and compare with prices for most other bulbous plants. In fact, thousands of different varieties, types, colors, and styles have been developed. No doubt, most are descended in one way or another from the tulip that perhaps is the rose of the Scriptures. Personally, I am not fully convinced that the tulip is what is meant in the Scripture at the beginning of this section. However, I like tulips, and because many Biblical gardeners believe they are indeed meant in this passage, I'll continue adding them to my home landscape.

One of my favorites to resemble this native tulip is a praetans called Fusilier. It matures 8 to 10 inches high in Zones 3 to 8. A vivid early spring flowering tulip, it fits rock gardens, nooks, and crannies; naturalizes well; and produces 5 to 6 flowers on each stem, which is unusual and gives massed color when spaced 4 inches apart.

Botanical wild tulips from Dutch Gardens mail-order firm also offer low-growing size. A collection of red and yellow colors provide a pleasing contrast and bloom mid to late spring only 4 to 6 inches tall with open blossoms. They naturalize well for years of perennial beauty.

Typically, tulips are erect plants with long, broad leaves and cup-shaped, solitary flowers at the tip of the stem. A taller favorite of mine is Red Emperor, one of America's best-loved tulips. It produces large, bright red blooms about 14 to 16 inches tall. Space 4 inches apart in Zones 3 to 8 for early spring flowering. Red Riding Hood and Scheherazade are others that provide a dramatic red color mass if you wish.

Many new hybrids offer double and even peony-type blooms. Best-known varieties include the Darwin tulip, a late-flowering type with tall, strong stems. The parrot tulip is another late-flowering tip with wrinkled petals. Early flowering and mid-season ones are included in mail-order catalogs. Choosing a variety of types today enables you to spread the tulip-blooming season over more weeks.

Look though mail-order catalogs each summer, from the list you'll find in Chapter 11. Dozens of firms offer a wide range of all types of tulips to delight every gardener. Red-flowered, low-growing tulips look most like those that still grace wild areas in the Holy Land and probably are a more appropriate addition to Biblical gardens.

Greigii species and hybrids are other attractive types of tulips. They offer you large flowers as well as attractive foliage. Related to the wild Oriental Greigii tulips, these plants do well in good sun or semi-sun. Species tulips are particularly valuable because they multiply and require little care. They bloom very early and are attractive in rock gardens, along walls and walks, or in naturalized settings. Praestans tulips seem remarkably close to the wild tulip of the Holy Land as well. Princeps have a brilliant scarlet color for dramatic displays. You may, of course, elect to plant the Kaufmanniana lily-flowered tulip, Fosterana hybrids, or even fancier varieties. Check out different types of tulips in colorful mail-order catalogs. There are so many different types and varieties you will be astounded and delighted with the wide choices you have. Remember that the simpler ones more closely resemble the actual tulips of the Holy Land.

Growing Tips: Plant tulip bulbs between early October and mid-December. If you plant too early, they may begin active growth in warm fall weather and can suffer a setback during winter. It is better to wait until you have planted your narcissi before setting tulips. They should be in the ground well before it normally freezes in your area if they are to set firm roots for strong spring growth. Your best advice for planting time is your county agricultural agent, a state-federal employee, part of whose job it is to provide residents with good local gardening information. Another key source is your local nursery or garden center.

Ideal soil is a light, fertile, well-drained loam. Never use fresh manure on tulip beds because tulip bulbs are sensitive. Well-rooted compost made with manure, weeds, and other organic material is a good soil additive to improve growing conditions. Most tulips produce their best blooms the spring following planting. Some varieties actually bloom well for several years, but others can decline. Adding fertilizer in spring and after blooming, using granular slow-release pellets or liquid spray-on fertilizer, can benefit tulips. Species tulips are hardier and tend to multiply themselves to add to your garden's beauty year after year.

As with narcissi and hyacinths, leave tulip foliage alone until it finally yellows. Foliage is the food-building part of these plants. After tulips bloom, the leaves must make and store food within the bulb for next year's performance. In cold northern areas, it pays to mulch over tulip areas in your gardens with straw, old dried leaves, peat moss, or compost. In spring, remove the top layer of mulch as your garden plants begin to sprout. Compost can be carefully cultivated into the soil as a beneficial conditioner.

Gardeners in southern areas have worried that they can't enjoy tulips as well as those in northern climates where these plants thrive. By refrigerating tulip bulbs for several months, you can induce a "winter-like" dormancy in the bulbs. Planting in December to early February should reward you with growth and bloom equal to those found anywhere else.

Chapter 3

℘

Flowers of the Holy Land, Growing Wild and Wonderful

Flower-blooming time is short in the rainy, growing season of the Holy Land. But what a glorious colorful display occurs from February into April. This is a time of dramatic change. After the dry, dormant time, the growing season virtually explodes with a profusion of bright colors in fields, meadows, rocky places, and along roads and byways. Many are mentioned directly in the Scriptures. You'll find the flowers for which we can give Scriptural references in the preceding chapter. Those include anemone, crocus, cyclamen, hyacinth, iris, lily, star-of-Bethlehem, and tulip.

Many other flowers are not clearly identified in the Bible and are indicated by words and phrases in the Bible like "flowers of the field," "lilies of the field," and other less specific references to the flora of the Holy Land in Biblical times. For example, from St. Luke 12:28: "If then God so clothe the grass, which is today in the field, and tomorrow is cast into the oven; how much more will he clothe you, O ye of little faith?"

My focus has been on those species that provide the best blooming beauty, are reasonably easy to grow, are suitable for gardens in North America and Europe, and are available to most home gardeners. With those as my guidelines, I've harvested information about them and provided the best growing tips, as appropriate for Biblical gardens.

You'll find some flower lists from key Biblical gardens in this chapter. I've condensed the lists from the many plants that grow in these gardens to just the flowers. Where these sharing gardeners have provided suggested Scriptural references, I have also included those too. The Biblical flower lists at the end of this chapter have been provided by dedicated gardeners of abiding faith. With thanks to those who have provided their lists of plants that grow in their church, temple, or private gardens, I have compiled my composite list. I have included specific references here to varieties that are most readily available. Where some flowers are difficult to find in a type most similar to those that grow today in the Holy Land, I've also provided appropriate, similar species of the same genus. I gratefully acknowledge the extensive research by other gardeners and scholars as they methodically searched and planned and planted their gardens. Although at times I do not quite agree or accept some of those choices, one major consideration rises to the top. God has given us the beauty of flowers in the Holy Land and throughout the world. With this simple thought, you are free to choose among the flowers included in this chapter.

You'll find details about many wonderful plants along with practical how-to ideas, tips and advice for growing them successfully. I have included this key list with three basic considerations in mind. These plants do indeed have their roots in the Holy Land. They also can be grown in gardens in most of the United States, or at least related ones can be grown successfully. In each specific flower section I've suggested varieties that I believe would be appropriate and worthwhile. Finally, they are most likely to have been mentioned in the Scriptures in generic terms as lilies, grasses, herbs, or otherwise. At the end of this chapter you'll find lists of flowers that grow in some of the most extensive Biblical gardens in America which I've included for reference. With those considerations, take your pick and grow productively.

Other Biblical Flowers of the Holy Land

Flower Name	*Scriptural Reference*	*Latin Name*
Buttercup	Psalms 4:5	*Ranunculus*
Chrysanthemum	Various	*Chrysanthemum coronarium L.*
Daisy, Shasta	1 Kings 6:29, Psalm 45	*Chrysanthemum sp.*
Dandelion	Exodus 12:8	*Taraxacum officinale*
Flax, Blue Perennial	John 19:40, Proverbs 31:13	*Linum perenne*
Hibiscus	James 1:9–10, Job 30:4	*Hisbiscus syriacus*
Hollyhock	Job 6:6–7	*Alcea setosa*
Iris, Blue Flag	Hosea 14:5	*Iris versicolor*
Iris, Purple	Hosea 14:5	*Iris astropupurea*
Iris, Yellow Flag	Hosea 14:5	*Iris pseudacorus*
Larkspur, Delphinium	Luke 12:27, Song of Solomon 6:24	*Delphinium ajacis*
Lavender	Song of Solomon 6:24	*Lavendula stoechas*
Lenten Rose	Song of Solomon 6:24	*Helleborus*
Lily, Water	1 Kings 7:19, 1 Kings 7:22	*Nymphaea lotus—white, Nymphaea caerulea—blue*
Lupine	James 1:9–10	*Lupinus augustifolius*
Loosestrife	Song of Solomon 1:14	*Lythrum salicaria*
Mallow	Job 6:6	*Malva moschata*
Poppy, Common	Isaiah 40:6, Jeremiah 8:14	*Papaver rhoeas*
Poppy, Icelandic	Isaiah 40:8	*Papaver alpinum*
Poppy, Oriental	Isaiah 40:6	*Papaver orientale*
Red Buttercup	1 Peter 1:24	*Ranunculus asiaticus*
Rockrose	Isaiah 35:1	*Cistus sp., Cistus creticus L.*
Rose, Rugosa	Isaiah 35:1, Wisdom 24:14	*Rosa rugosa*
Phoenicia Rose	Ecclesiasticus 24:14	*Rosa phoenicia*
Salvia—Sage	Exodus 25:31–33	*Salvia officinalis*
Globe Thistle	Judges 8:16	*Echinops viscosus DC.*
Violet	James 1:10, Song of Songs	*Viola odorata*

Planting, Care, and Cultivation
of Holy Land Flowers

BUTTERCUP—*RANUNCULUS ASIATICUS*

Scarlet crowfoot or red buttercup is *Ranunculus asiaticus,* and is one of the more dramatic flowers that grows wild in the Holy Land. Although most yellow buttercups prefer moist areas, this unique red buttercup survives well into the dry areas. This plant typically blooms after anemones in its native environment and provides a great abundance of large red flowers from short stems that also bear the foliage. It qualifies as one of the "flowers of the field" in the Scriptures.

Unlike anemones which grow from rhizomes, the red buttercup produces thick storage roots alongside its fibrous absorption roots. Botanists say these change annually, which indicates that the plant uses its stored food reserves to send up its blooms. The leaves then produce new roots, much like bulbous plants rebuild next year's flowers from this year's leaves as they also store nutrients for the future growing season back in the bulbs. Unlike the tall yellow buttercups so common in pastures, meadows, and roadside ditches around the world, this red buttercup is less known elsewhere in the world. It may be difficult to obtain, which would require a substitute from the Ranunculus family. Buttercups make up the family Ranunculaceae of the order Ranunculales. The tall buttercup so common as a wildflower in America is classified as Ranunculus acris.

Plant Profile: Short stems produce both flower buds and foliage. Lower leaves are shallowly lobed but higher leaves have toothed lobes. The scape branches above and terminates in typical single crimson flowers with five sepals and five petals. Productive and profuse in its native Holy Land habitat. Tall common yellow buttercups sprout from fibrous roots to become 2- to 3-foot-tall plants. Prefers sun or shady areas and can persist for years. May be somewhat invasive.

Growing Tips: Plant Persian ranunculus, *R. Asiaticus* tubers in fall for bloom in early spring in moderate climates. Prefers full sun, well-drained, good soil. May not thrive in cold zones, but you can bring pot indoors for the winter and replant next season. Popular strain is Tecolote Giants in single and mixed colors. More common yellow buttercups are available from some mail-order firms or you can try transplanting them from wild locations. Dig well around them to obtain sufficient root system with soil for successful transplanting. These plants also can be invasive, so curtail their spread if necessary. Planting in containers is one solution.

CHAMOMILE—*CHAMAEMELUM NOBILE*

This oldtime plant has small white daisy flowers with yellow button centers and emerald green leaves that are scented of apple and camphor. A delight along pathways where bruising them releases their sweet fragrance. This was a colonial herb for repelling insects and has been widely employed medicinally. The yellow-white, fairly large flower heads of dog chamomile really catch your eye when they are blooming. Many scholars agree that this must be considered among the "flowers of the field." No specific name is given to the species in Scripture, but there are 150 species in the Composite or daisy family, most in the Eastern Mediterrean area. Most are annuals with self-seeding ability. Chamomile blooms have the unique habit of turning the peripheral flowers down in the evening and spreading them out in the morning.

Plant Profile: Chamomile is a low-growing perennial for Zones 4 to 8 that matures about 12 inches tall. Some types grow up to 3 feet high. They like warm areas and dry, sandy soils, and the taller types may need staking if your garden is subject to wind gusts.

Growing Tips: Plant in well-drained soil according to directions on the seed package. Likes full sun or partial shade and moderate moisture. This evergreen perennial forms a spreading 6- to 12-inch mat of bright light green aromatic leaves. Plant divisions 1 foot apart. Can be used as a ground cover and mowed periodically to keep under control.

CHRYSANTHEMUM—*CHRYSANTHEMUM CORONARIUM*—CROWN DAISIES

Crown daisies, and related corn marigolds, have long grown wild along roadsides, field margins, and elsewhere in the Holy Land. Some may consider them weeds, but they are glorious golden ones, no doubt. Perhaps they, too, are among the "flowers of the field" mentioned in Isaiah, James, and Peter, since they do indeed flower and fade, just as human life passes away. These flowers reach 20 to 30 inches tall and have finely divided leaves and yellow flowers in summer. Chrysanthemum is the common name for numerous perennials. They belong to the family Compositae and about 100 species are classified in the genus Chrysanthemum. The actual chrysanthemum found in fields in the Holy Land is *C. coronarium*, which is often called crown daisy, a delightful yellow bloom that reminds me of golden coreopsis. Today, we may think of chrysanthemums as the hybridized florists' mums with large, globular heads in which ray fields are greatly multiplied. Garden chrysanthemums that more closely resemble those growing native in the Holy Land are easily available and grow well in outdoor gardens.

My Biblical gardening friend, Rev. Marsh Hudson-Knapp, had some thoughtful observations from the New Testament about "flowers of the field." He notes that Jesus drew his followers' attention to the plants that bloomed abundantly around him as signs of God's abiding care. "And why are you anxious?" he asked. "Consider the lilies of the field, they neither toil nor spin, yet even Solomon in all his glory was not arrayed like one of these." For his Biblical gardens at the First Congregational Church in Fair Haven, Vermont, Rev. Hudson-Knapp points out, "Our Shasta daisies are among the many cousins of plants that would have been 'lilies of the field' in the Holy Land, constantly reminding people of God's unending care for us. Chrysanthemums, crown anemones, ranunculus or wind flowers all grace the Bible's natural world. So also did delphinium (larkspur), hibiscus, and lupine.

"The New Testament reminds us not only of our blessings, but also of our responsibilities. The first Letter of Peter urges us to deeply root ourselves in God and God's word. In time, 'The grass withers and the flower falls, but the Word of the Lord abides forever.'" (I Peter 1:24–25).

Rev. Hudson-Knapp goes on to explain many other Biblical messages that come alive as we reflect on the Biblical background of our plants. Each plant has its special significance, both from Scriptural writings and from its roots in the traditions of the Holy Land. His garden is one of the most comprehensive Biblical gardens to see, a bit small in actual size but numerous in variety of plants and huge in its worthwhile message. It also serves as a fine example for all who wish to begin their own authentic and symbolic gardens at their homes, churches or temples, or other cooperating organizations.

Crown daisy, *C. coronarium*, was introduced from the Mediterranean to England before 1630 according to old records. Supposedly the flower's name comes from the Greek word *chrysos*, which means "golden," and *anthemon*, which means "flower." This term was used by early Greeks to identify most golden yellow flowers. Crown daisies grow abundantly in Israel today, and in all likelihood fit the passage from James 1:9–10, "Let the lowly brother boast in his exaltation and the rich in his humiliation, because like the flower of the grass he will pass away."

By the 1700s, doubled types of chrysanthemums were bred and became popular. Both types were brought to America by colonists. Crown daisies, their early name, bloom from midsummer to late fall with attractive yellow flowers and grow 2 feet tall.

C. coronarium is the crown daisy. *C. segetum* is the correct term for a similar common daisy, better known as the corn marigold. It is shorter, 8 to 15 inches high.

Plant Profile: This common chrysanthemum, crown daisy is an annual plant with branching, leafy stems. Yellow flower heads are 1½ to 2 inches across with many florets. Plants grow 1 to 2 feet tall, depending on growing conditions, and become attractive bushy displays. When pollinated by insects, the ray florets tend to bend over and press themslves into the head.

Growing Tips: Plant according to seed packet directions in well-drained soil. Because it has a tradition of growing well in wastes, along roadsides, and in other poor soil areas, the crown daisy has a reputation of easy growth and profuse golden blooms. Sow seeds inside in deep pots or outside in late spring. Transplant or thin to 4 inches apart. Keep soil moist until they are well established. If you don't have the patience to make sure your seedlings never dry out, purchasing plants may be wise. They will fill your garden and show off their blooms more quickly. Open, sunny locations are important and they prefer light, sandy loam. By snipping off spent flowers you save the plant's energy expended trying to make seeds and thereby force more flowers to bloom.

DAISY, SHASTA—*CHRYSANTHEMUM LEUCANTHEMUM*

Daisies and daisy-like plants are noted for their abundance in the Holy Land and many countries around the world. Many scholars believe they are included in the collective term "flowers of the field" and I concur. As the Scriptures note, "All flesh is grass, and all its beauty is like the flower of the field. The grass withers, the flower fades, but the word of our God will stand forever" (Isaiah 40:6 and I Peter 1:24–25). Grass combined with flowers in this and other verses in the Bible most likely indicates flowers of some sort, as in James 1:10, "like the flower of the grass he will pass away," which reminds us of a short, duration of our human lives. Matthew 6:30, writing about the "grass of the field," is also more likely a reference to flowers than merely green grass. Daisy is the common name for a number of flowering herbs of a family of composite flowers. True daisies include the old-fashioned English daisy and the Shasta daisy. Other species include the multicolored African daisy and the black-eyed Susan, a common wildflower in the United States with yellow rays and a dark brown disk.

Ox-eye daisies are a traditional American wildflower and happily hardy. Originally from Europe, they are truly naturalized Americans and grow well all across the country. Fern-like leaves give plants a graceful look. Sparkling white petals with golden eyes bloom from April to August, depending on where you live. Plants stand 15 to 24 inches tall on sturdy stems. Flowers are 1 inch to nearly 2 inches across and ideal for cutting. Daisies long have been praised in poem and song. Young children have traditionally used them to make daisy chains and necklaces. Young lovers have pulled the petals one by one, saying, "He loves me, he loves me not; she loves me, she loves me not."

Daisies are easy to grow and surprisingly prolific. They love sun. Best planting time is spring or fall. From wild plants, select and divide large clumps. Keep roots wrapped in moist sphagnum moss covered with plastic until you plant them. You can also grow wild ox-eye daisies, *C. leucanthemum vulgare,* by collecting seeds in late summer from mature plants. Sow them in the fall, as nature would, or save for spring planting. Shasta daisies, *C. maximum* and *Leucanthemum superbum*, are a super summer and fall bloomer. Original types have been replaced by exceptional hybrids for better-formed, longer-blooming plants. Biblical gardeners rightfully prefer Shasta daisies to wild ones for their larger, more dramatic blooms.

Plant Profile: Cultivated Shasta daisies have pure white, single daisies with long, strong stems, blossoms 3 to 5 inches across, depending on modern hybrid variety. One of the most useful of all perennials. Typically grows 2 to 3 feet tall, extremely hardy. Thrives in Zones 4 to 9.

Growing Tips: Plant seeds as directed or set started plants 1 foot apart in deep rich soil in a sunny location. Remove spent blooms to extend flowering season. Shastas can outgrow their spot, so divide every few years. Alaska variety blooms early with 4-inch flowers. Becky performs well even in the South. Snowcap is a Wayside Gardens introduction with long-lasting flowers that are weather resistant and produce abundant blooms in mid-summer on mounded 12- to 18-inch-high plants. Set out divisions of Shasta daisies in early spring or fall and container-grown plants anytime. They thrive in sun, but do well in partial shade in hot summer areas. In cold areas, mulch around plants for winter protection.

DANDELION—*TARAXACUM OFFICINALE*

"They shall eat the flesh that night, roasted; with unleavened bread and bitter herbs they shall eat it" (Exodus 12:8). In this passage, various bitter herbs could be considered, including dandelions, according to both Biblical and botanical scholars. The poppy-leaved Reichardia is a desert plant with a rosette of lobed leaves, with large flowering heads that closely resemble dandelions. Because that flowering herb, the dandelion, is so widely grown as a food in Mediterranean countries and is widely known around the world, dandelions deserve recognition at last. Unfortunately, in America they are considered a weed in otherwise beautiful green lawns. Eradication seems to be the order of the day, every year. Nevertheless dandelions are a plant with Holy Land roots and deserve to be considered for Biblical gardens. Of note, the dandelion's perserverence and resiliency makes it the favorite Easter flower for Rev. Marsh Hudson-Knapp. He brings dandelions into worship for the children as a reminder of God's indefeatable resurrection power!

Each year, I watch thousands of dandelions sprout and display their golden blossoms in my back lawn and meadow. They last only a few brief weeks, but few other flowers can provide such a profuse display. It seems sad that Americans feel compelled to follow the dictates of chemical companies and try to eradicate dandelions. Each year we spread uncounted tons of poisonous herbicides on lawns to kill weeds. How much of those pesticides seeps into groundwater, wells, and streams is never told. Perhaps it is best not to know how badly we are poisoning our planet, just trying to have perfect green lawns. It does seem ironic. Today, we focus on eating natural foods, growing tastier vegetables organically, and worrying about the environment. Somehow, we neglect the environment in our own lawns as tons of herbicides are used nationwide to eradicate dandelions every year. Yet, back they come. Their determination qualifies them for a place in Biblical gardens, or perhaps a few left to grow peacefully along the edge of a lawn.

In many European countries, dandelions are favored for salad greens and are also a cooked green, much like spinach. Flowers are often used for making wine. Actually, dandelions are cultivated in Europe and the roots are ground up as a coffee substitute. Strange as it may seem with America's compulsion to rid the country of dandelions, several American garden mail-order companies offer dandelion seeds, for those who wish to grow them as food, rather than think of dandelions only as a pesky weed. Johnny's offers Italian dandelion, *Cichorium intybus*, which is not a true dandelion but a blue-flowered chicory. Typical yellow wild dandelions are much earlier, upright growing, larger and deeper green. There are several varieties in Italy, where the leaves and flower stalks are both used in salad.

Plant Profile: Dandelion is the common name for a stemless perennial herb of the composite flower family. The species has long taproots, rosettes of lance-like leaves, and a flat flower head of bright golden florets borne on hollow, stem-like stalks.

Growing Tips: To grow dandelions, start as you would with lettuce by direct seeding anytime during the season. Space plants 6 to 8 inches apart and clip leaves before bloom for less bitter taste, usually when they are less than 10 inches tall. Eventually the plants will make a flower stalk, but with characteristic blue chicory flowers if you plant the typical Italian type.

Catalogna special is the finest Italian dandelion, according to Johnny's specialists. Naturally, most of us think in terms of the typical yellow dandelion that adorns lawns, fields, and roadsides and seems to grow easily and prolifically everywhere, especially where we don't want it to grow. For the sake of species integrity in a Biblical plant garden, a few dandelions deserve their spot. Simply pick a few seed heads and bury them in the desired spot in your garden where they'll be one of the first flowers to bloom in the spring. To prevent their spreading, simply clip off the flower heads when they puff up to seeds. You may dig up a few as a service to your neighbors. Use a long shovel and dig only young plants because even they can have deep tap roots. They may wonder why anyone wants to grow dandelions. That will give you an excellent opportunity to open the exciting and meaningful topic of Biblical gardens to encourage more Biblical gardens around your neighborhood, even if they won't grow dandelions!

FLAX, BLUE PERENNIAL—*LINUM PERENNE*

In Exodus 9:31, as we read about the effects of the hail sent by God as one of the ten plagues on Egypt, we find, "The flax and the barley were ruined, for the barley was in the ear and the flax was in the bud." Flax was a very important crop to the Egyptians. Much later, in John 19:40 we read, "They took the body of Jesus, and bound it in linen clothes with the spices, as is the burial custom of the Jews." Linen, of course, is made from flax. It is one of the world's oldest clothing materials. The flax plant is one of about 200 species of the genus Linum and has been grown for its fibers from time immemorial. Ancient writings record flax cultivation for several millennia before the birth of Christ. Flax fiber for cloth originated more than 10,000 years ago. Pieces of linen fishing nets and clothing have been found in Stone Age lake dwellings. Ancient Egyptians used linen shrouds and passages in the Bible mention the manufacture of linen. Workers soaked and beat the flax stems to soften them and remove the outside. The strong inner fibers were used for weaving the linen. Some museums still demonstrate this age-old practice today, growing the flax and processing the stems and finally weaving linen. It is a remarkable process.

Flax is the common name for a family of plants. One species is grown extensively for its fiber, which makes linen, and seed, which makes linseed oil. For millennia, flax fiber products included linen threads and fabrics and was widely used by the

Egyptians. Blue meadow flax, *Linum perenne*, was introduced by Johnny's in 2000. It is an airy flower that welcomes spring with waves of billowing blue. Following a month of blooms, decorative round seed pods form. It can be cut back for later rebloom. This flax sows itself and plants are 16 to 24 inches tall.

Plant Profile: Cultivated flax is an annual herb of 1 to 2 feet tall with erect stem branches toward the top. Branches have long narrow leaves with showy blue flowers that have five sepals, five petals, and five stamens. Flax plants have shallow taproots. Because the stems contain the fiber, the taller varieties, which are sparsely branched, are mainly used. Although there are various colors, the truest to tradition is the blue flax. Other types include alpine flax, *L. alpinum* and *L. usitatissimum*.

Growing Tips: Plant seeds somewhat thickly in ordinary garden soil in spring. They grow rather quickly as a compact bed or group. Flowers appear in two to three months when plants are 15 to 18 inches tall. Blooms soon fall, but this plant provides a succession of blossoms for several weeks from other buds that form. Keep in mind, flax doesn't like cold, soggy ground. It prefers sunny locations as its native habitat provided.

HIBISCUS—*HIBISCUS SYRIACUS*

Several types of hibiscus are grown today for their large and showy flowers. Hibiscus is a genus of plants, commonly called rose mallows in the mallow family. It is native to warm, temperate regions of the northern hemisphere. The genus Hibiscus belongs to the family Malvaceae. There has been much debate about the true meaning of the word "rose" in some Biblical translations. No doubt, there are wild roses growing in the Holy Land today as there were 2000 years ago. You'll find details about them and suggestions for matching these wild, native roses in their part of this chapter. Here I focus on hibiscus. The rose of Sharon is a member of the hibiscus group. It is a deciduous Asian shrub that grows 8 to 12 feet tall, is upright and compact while young but spreads with age. Cultivated varieties have open, bell-shaped flowers in a variety of colors. In reality, it resembles a bush covered with blooms that look like hollyhock flowers from mid to late summer, often up to frost.

Blossoms are single to double, depending on the breeding of new varieties. The rose-mallow, a tall perennial herb with ovate leaves and white flowers with rose centers, grows in salt marshes of the eastern United States. Other types of mallows are included under "Mallows" in this chapter. There are other types of hibiscus, from the red-leaf type, *H. acetosella*, grown from seed as an annual in northern areas, to the Chinese hibiscus, *H. rosa-sinensis*, which is a tender evergreen shrub. Because of the wide choice, you are well advised to carefully consult catalogs with specific cultural tips for various types before you add it to your outdoor garden list.

Some veteran Biblical gardeners prefer the rose of Sharon, *H. syriacus*, as a display and specimen plant in their gardens. They have more than one reason, no doubt. Some equate that plant with the word "rose" in Scriptures. Most scholars disagree but make your own choice. No doubt, rose of Sharon bushes are a delightful, profusely blooming garden delight.

Plant Profile: Hibiscus range from annuals to perennials, depending on the type. Most grow from 5 to 8 feet. Because of the wide choice among versions of this plant, it is best to check with catalogs to obtain details for the type that seems best for your intended uses, whether annuals or perennial growth habit.

Growing Tips: Chinese hibiscus, also known as tropical hibiscus, is one of the showiest plants, a tender evergreen that thrives in summer outdoors or in containers on patios, but must be indoors during winter. Requires excellent drainage. Improve soil with soil mix and compost. Fertilize monthly April to September outdoors, then bring in before frost. Prune to desired shapes. Pinching tips in spring and summer increases flower formation. Another type of hibiscus, the rose of Sharon, is easy to grow. This is a deciduous perennial shrub that prefers heat and tolerates drought. You can prune it to desired shape for bigger flowers or to keep it within bounds. It pays to protect young plants with a winter ground mulch for the first few years. Some recommended varieties include Ardens with double purple flower, Blue bird with single blue blooms and a red eye, Lucy with double deep rose flowers, and many more. Enjoy catalog window shopping.

HOLLYHOCK—ALCEA—*ALTHAEA ROSEA*

Biblical scholars have considered the Hebrew word *halamuth* in the original passage from Job 6:6–7, "Can that which is tasteless be eaten without salt, or is there any taste in the slime of the purslane?" Scholarly thought is that this passage refers to one or more species of the genera Malva and Alcea, both in the Malvaceae family and common in the land. If it is *Malva sylvestris*, it would be a mallow. If not, it might be *Alcea setosa*, a hollyhock.

Hollyhock is the common name of a genus Alcea of the family Malvaceae, of herbs cultivated in gardens of Europe and the United States. We know them best as tall, erect, single-stem 5- to 9-feet-high plants with heart-shaped, wrinkled, and hairy leaves. Oldtime hollyhock varieties bore single flowers in a variety of colors. Modern hybridizers have perfected gorgeous double blooms in white, yellow, salmon, rose, red, violet, and pastel shades too. No doubt they are making a major comeback in gardens for their classic beauty, tall growth habit, and colorful displays. These plants were and are found in the fields of the Holy Land.

According to folklore, original wild species of Hollyhock were originally called Holy Hocks by Crusaders. This was probably *Alcea setosa*. Because it is rare beyond the Mediterranean, *Althaea rosea* has been used by Biblical gardeners. An ongoing problem is that hollyhocks are often susceptible to rust and insect damage. Yet once deeply rooted, they can last for years. As proof, note old hollyhocks around the foundations of deserted farmhouses in the countryside. Mail-order catalogs offer both the single-flower form and hybridized double-bloom types. The single-flower type most closely represent the originals of aeons past.

Plant Profile: A favorite flower since medieval times, it reputedly arrived in England with returning Crusaders. Hollyhocks were brought to North America by colonists. Traditionally a tall background plant used along house and barn walls, hollyhocks grow 6 to 8 feet tall. Best used in groups or masses for most effective display. The original species, called Antwerp hollyhock, was yellow, according to legend. A yellow saucershaped rare hollyhock still is available from Select Seeds Antique Flowers as started plants. It blooms from late June to September. New hybrids provide many colors. *Alcea ficifolia* is a true perennial that persists for years and is less susceptible to rust disease. *A. rugosa*, a rare species of traditional yellow hollyhocks, is offered as plants by Select Seeds. Hollyhocks grow well in Zones 3 to 10. An All American winner in 1939, Indian Spring has semidouble flowers of white, soft pink, and deep rose. It often blooms the first summer if sown indoors to start seedlings. This is a biennial, as many others are. Check catalogs to be sure you order perennial or biennial as you prefer. Johnny's elegant old-fashioned single mix grows in Zones 3 to 10, maturing 48 to 72 inches tall in fertile soil rich in humus, but prefers cool summers and locations out of the wind. Chatter's double mix is frilly, fully double blooms in romantic colors. An heirloom introduced about 1880, it is back again in golden yellow, maroon, pink, scarlet, and white. Plant in consecutive years to have a self-seeding perennial hollyhock border. Another option is miniature hollyhock, *Sidalcea Elsie Heugh*. It bears soft pink flowers on 24- to 30-inch-high plants and is carefree when planted in a sunny or lightly shaded spot in normal soil. Good for bouquets too.

Growing Tips: Plant seeds in the ground in the fall, August through September, for next year's blooms if biennials. Follow directions on seed packages for the type you buy. Pre–spring planting in peat pots or containers gives hollyhocks a good indoor start for spring planting. When weather is warm, carefully transplant to their permanent outdoor home. Some types are true perennials and can be started from seeds or bought as plants. Select Seeds has both types and features beautiful rare ones.

LARKSPUR, DELPHINIUM—*DELPHINIUM AJACIS*

Larkspur is the common name for members of a genus of flowers in the buttercup family. Flowers may be purple to pink or white and grow in loose clusters with feathery-appearing leaves. Larkspur is often used interchangeably with the name of delphinium by many people. Although that is not technically correct to purists, for our purpose we focus on larkspur as a plant of the Holy Land aeons ago and still growing colorfully today.

Giant Imperial series larkspur, *Consolida ambigua*, has colorful, old-fashioned spires. Plants grow 30 to 48 inches tall and prefer cool summers and rich, well-drained and slightly alkaline soil in full sun. This lovely flower retains its brilliant colors even when dried. Be aware that all parts of larkspur are poisonous, including seeds. Exercise extreme caution around children and pets. Johnny's offers delphinium as a separate plant, *D. belladonna*, a perfect blue flower for any garden. Free flowering, deep blue to icy white. Plants started in April will bloom in late August to killing frost. The second year and thereafter they bloom June to July and again in August if cut back. Good in Zones 3 to 7. This plant, too, has poisonous parts.

Plant Profile: These dramatic plants produce elegant spire-like blooms that attract all types of birds. They are effective in background borders for their 2- to 3-feet-tall displays. Larkspur Blue Bell, *Consolida ambigua*, has blue spires in spring and early summer. This antique variety won All American Selection honors back in 1934. Blue Cloud, *C. regals*, has a bushy cloud of blooms and also grows 3 feet tall. Larkspur Earl Grey, *C. ambigua*, dates back to their welcome in England in 1572. Other types are available from various seed firms. A bonus with larkspurs, they attract butterflies and hummingbirds to their blooms. Keep away from children and pets, of course, because of toxins in the plants.

Growing Tips: Old-fashioned larkspurs are favorite cut flowers. As cool weather plants they must have cold weather during germination and seedling growth. Sow four weeks before the last spring frost in northern areas. In southern areas, sow in October and November. Seeds take a while to germinate so early planting is important. Provide well-drained soil in sun to part shade. Thin for stronger, better-blooming plants. These combine well with Shirley poppies and pansies as cool weather plants. Larkspurs will self-sow to become carefree flower members in your gardens. Grow in Zones 1 to 9.

LAVENDER—*LAVENDULA STOECHAS*

Lavender is the common name for a genus of fragrant herbs native to Eurasia. The flowers are borne on terminal spikes and modern varieties produce profuse blooms. One new introduction, Lavender Lady, has been honored as an All America selection. Common lavender is a shrubby Mediterranean herb also cultivated in gardens. It has narrow leaves and small lilac purple flowers containing oil of lavender, used in perfumes and toilet water. Dried flowers of lavender are often popular as sachets for perfuming clothing.

In addition to the oldtime English lavender, *L. augustifolia*, which has pleased generations of gardeners, lavender is now available in many other attractive and distinctive types. Lavender Hidcote is a cultivar with rich, deep purple flowers and gray-green needle-like leaves. It is more compact in growth habit. Lavender Grosso, *x intermedia,* arose as a cross of English lavender and spike lavender, back in the 1820s. Today French fields are filled with this lavender which supplies much of the French oil of lavandein used in perfumes and soap making. Large, sweetly pungent spikes of deep purple, 3 to 6 inches long, wave above the gray-green foliage in summer. It thrives in Zones 6 to 10.

A Spanish lavender, *L. stoechas,* has small leaves that possess a scent akin to rosemary. The deep violet, velveteen flowers are topped with ¾-inch-long petal-like bracts. It blooms almost continuously, making it a superb potted flower 2 to 3 feet tall and fine in Zones 7 to 10, according to specialists at Select Seeds, an antique flower specialist firm.

Plant Profile: English lavender, *Lavandula augustifolia*, is an antique plant. The Latin means "to wash" and indicates the long history of this plant as a beautifying, soothing, antiseptic agent. It is an essential in potpourris. Perennial plants grow 2 feet tall in horticultural Zones 5 to 9.

Growing Tips: Plant in well-drained soil in early spring according to directions on the package, or use prestarted container plants. Lavender is somewhat drought tolerant. Cut back in early spring for more profuse blooms from this perennial.

LENTEN ROSE—*HELLEBORUS*

The legendary Lenten Rose, which traces its roots to Scripture and the soil of the Holy Land, has been rediscovered, or perhaps more accurately, reintroduced. After fifty years of dedicated breeding work and ten years of collection and ongoing selection, John Elsley, the Director of Horticulture at Wayside Gardens, has developed a replica Lenten rose. A premier introduction as Royal Heritage Hellebores, Elsley believes these are the finest new perennial their garden firm has introduced in a decade. Even better, these special Royal Heritage Hellebores, roses, are candidates to replace hostas as a carefree perennial for shade and partial sun areas. This magnificent new hellebore blooms in late winter and earliest spring when most of the garden is dormant. That's a blessing for all gardeners, welcome flower blooms even before crocuses appear! These plants have lovely palm-like foliage, form a clump that is attractive year round even under shrubs or shady areas. These delightful plants have 2-inch flowers with overlapping petals, reminiscent of wild roses one finds in America's meadows and roadside areas. In tests this strain produces flowers for nearly five months, from winter though spring. Cut and floated, they make an excellent centerpiece that can last up to two weeks.

Test gardeners report that these are very permanent, low-maintenance, and virtually disease- and pest-free. They are easy to establish from vigorous container stock and should develop into sturdy, long-lived clumps for years of bloom.

Another potential choice for a Biblical rose, if that is how you wish to interpret some of the Scriptural translations, could be the Christmas rose. *Helleborus niger* is a hard-to-locate legendary plant much loved by gardeners worldwide. Now Wayside Gardens offers plants that grow 10 to 12 inches tall and bloom in winter when the garden is at its lowest ebb. In fact, with moderate winters, some plants oblige with Christmas-time blossoms. This helleborus has 1½-inch-wide flowers that open white, mature to dusty pink, and last for months. The evergreen, leathery leaves provide bold texture and color in landscapes.

Plant Profile: Sturdy perennials are 18 to 24 inches high with blooms 2+ inches across in a variety of pinks, reds, some mottled, and white. Blooms are single petals which closely resemble the wild roses found even today in the Holy Land. Winter blooming habit makes this an eye-catching, conversation piece plant for any garden. Hardiness seems to be its middle name, according to gardeners who cherish that remarkable trait.

Growing Tips: Plant in well-drained, moisture-retentive soil with ample humus. Mulch helps preserve soil moisture and stop weeds as plants set permanent roots. Best for Zones 4 to 9. Plant clumps to mature 2 to 3 feet across. Tolerant of heat, humidity, and even periods of drought and can even thrive in clay soils.

LUPINE—*LUPINUS AUGUSTIFOLIUS*

Lupine is the common name for a large genus of annual or perennial herbs. These showy flowers, mostly violet or purple in their native wild habitats, are borne on long terminal spikes. The seed pods may be several inches long and look like dried pea pods on the lupine spikes after flowering. Lupines are one of the more delightful perennials in our wildflower banks and field areas. Once they set their tap roots, they appear again to grace the garden every spring. Some are white, pinkish, and even yellow in our native plant areas. Average height 30 to 36 inches. Zones 4 to 8. Lupines can be very hardy, long lived, and low maintenance. They thrive on ample moisture and do best in areas with cool summers. Wild lupines are often seen in masses in open fields and meadows, especially in New England states, where they have a firm roothold.

New hybrid varieties have been introduced that present spectacular multicolor displays each spring. Whether you wish the natural look or more dramatic color splashes of new varieties, lupines deserve a place in your outdoor living room. Note: Although a familiar sight growing wild and an attractive bloomer in gardens, be aware that all parts of this flowering plant are poisonous, including seeds. Exercise caution around children and pets. If you are ultracautious, simply omit these and other plants like larkspur from your garden. When buying any seeds or plants of any type, always ask your local garden center or nursery about them, and check in mail-order catalogs for any warnings about plants that you need to know.

Plant Profile: Lupine leaves are distinctive like the fingers on your hand and last beyond bloom to late summer as attractive foliage in the garden. In the wild, lupines are usually purple to white with shades in between. Flowers are sweet pea shaped on spikes at the ends of stems. New hybrids range across the rainbow from lemon yellow to red. Tap roots give lupines a base from which they can spread their seeds and colonies of plants around. The bank of our house is evidence of how well they self-propagate. Many hybrids offer you a wide and colorful choice. All resemble parents with roots in the Holy Land, much closer look-alikes than other plant hybrids we have seen and planted through the years.

Growing Tips: Tap roots are so deep it is nearly impossible to transplant lupines. Use seeds you buy or collect. Soak seeds overnight or scarify them with a metal file before planting. To naturalize in meadows or sod areas, use a shovel or bulb-planting tool to turn over earth, reinsert sod upside down, and plant lupine seeds on top with a ½-inch soil cover. Water well in spring or fall when you plant. Spring seeds may even produce plants that bloom come fall. In gardens, follow directions on seed packages. Lupines like sunlight and most accept poor soil conditions if they receive moderate water, especially when blooming. New hybrids prefer richer, slightly acidic, and well-drained soil. If you want a dazzling array of colors, the Van Bourgondien catalog features some of the best new lupine hybrids.

LOOSESTRIFE—*LYSIMACHIA PUNCTATA*

Loosestrife is a common name for the Lythraceae family of plants. Typical loosestrife genus is Lythrum and traces its roots to Eurasia. There are about thirty-eight native species there and in North America. The yellow ones most closely duplicate the wildings in the Holy Land. Popular yellow loosestrife is a June-to-August, long-season bloomer that adds color when spring blooms have passed by. These plants have a certain appeal but have been overlooked by many home gardeners, perhaps because this plant hasn't been promoted much by seed and plant firms. Because of its probably traditional roots as a wild plant in Palestine ages ago, it may be worth considering as an addition to your garden, although on a lower priority scale.

Whorled loosestrife, *Lysimachia punctata*, is a gardener's dream for problem areas of moist, shady spots where little else will grow. This herbaceous perennial reaches 24 to 30 inches tall, spreads to nearly 3 feet, and sports brown-throated 1-inch whorled blooms of clear yellow atop sturdy stalks. The whorled foliage reaches 3 inches long in clusters of three to four. If you like loosestrife and want a representative in your garden, this may be a solution. A warning! Avoid purple loosestrife, a related plant that is aggressively invasive. It can take over moist areas and crowd out natural ecosystem plants. Some states have begun banning the planting of it because of its invasive, dominant growing habit.

Plant Profile: Yellow loosestrife, *Lysimachia punctata*, is a useful, long-flowering, and easy-to-grow perennial choice. It has bright yellow flowers nestled in leaf axils, grows 3 to 4 feet tall, and is hardy in Zones 4 to 8.

Growing Tips: Follow seed package directions or use pre-started nursery plants. Loosestrifes love moisture but tolerate drier soil conditions in partial shade. They thrive near watersides and combine well with blue and yellow flag iris, and cattails if you wish to try them too.

MALLOW—*MALVA MOSCHATA, MALVA SYLVESTRIS*

Mallow is the common name for a family of 80 genera of herbs, shrubs, and some trees. Well-known members include the hollyhock, marsh mallow, and hibiscus. The floral whorls, calyx, and corolla are present in groups of five and the segments of sepals and petals are free from one another. Mallows make up the family Malvaceae of the order Malvales. The marsh mallow is *Althaea officinalis*. According to some researchers, *Malva sylvestris* is the mallow mentioned in Job 6:6.

We should recall that different translations of the Bible use different terms to interpret early Hebrew words into English. For example, the New English Bible calls this plant mallow. The Revised Standard Version names it purslane, but the New Revised Standard Version translates it as mallow. Other translations have their identification. That is a common problem in accurate plant identification from early translations through the present, so as Biblical gardeners we must decide which plants would be most representative and grow best in our areas. You'll find lists of flowers selected by some of America's best Biblical gardeners for their gardens on pages 90–95, as a guideline.

Whatever the translators have said, there is no doubt that mallow is found throughout the Holy Land as well as in North America and across Europe. It is a biennial plant growing 2 to 3 feet tall with pink or purple flowers. During recent garden tours here in Kennebunk we were delighted to see exceptional displays of mallow, colorful, abundant, and obviously thriving. It deserves a special display in more Biblical gardens. As we learn from the Bible, everyone had their times of trial, as we all do. One of the sufferings that Job experienced was loss of appetite.

At that time, hollyhock and mallow were used to flavor foods. To poor Job, however, as we read in Job 6:6–7, even the mallow had lost its taste.

The musk mallow, *H. moschata,* is a perennial erect branching plant growing 3 feet tall with finely cut leaves and flowers 1 to 2 inches across. These plants emit a mild, musky fragrance if foliage or blooms are bruised or crushed, hence its name. Rosea has pink flowers; short alba bears white flowers. Another mallow, *M. sylvestris*, is a perennial or biennial with flowers appearing all summer.

Plant Profile: Always at home in oldtime gardens, rose mallows, *M. moscheutos,* have sweet rose-colored blooms summer to fall. The large 6- to 7-inch saucer-shaped flowers are mostly pink with a purple eye. They bloom freely on plants that grow 4 to 5 feet tall. An excellent eye-catching plant for massing near a pond or waterway. Perennial in Zones 5 to 10.

Growing Tips: Plant from seed in rich garden soils. Add composted humus to give this perennial a good start for years of massed blooms. You may also wish to explore other mallow flower opportunities. Enjoy these new discoveries, and the wonders of heirloom varieties that are being rediscovered, brought back and now available to all of us from alert, dedicated specialty seed firms. You'll find some "flagged" in my favorite mail-order firm lists in Chapters 11 and 12.

POPPY, COMMON—*PAPAVER RHOEAS*

Poppies provide spring and summer delights . They are indeed one of the wild beauties of the Holy Land, sparkling in meadows and emerging among other plants every growing season there. Typically, poppy blooms are short-lived, lasting only a few days. Because they fade so quickly, many scholars believe this identifies them also with the context of the "flower of the grass" in I Peter 1:24—"All flesh is like grass and all its glory like the flower of the grass. The grass withers, and the flower falls, but the Word of the Lord abides forever." It also seems to fit the scene depicted in Isaiah 40:6, "All flesh is grass, and all its beauty is like the flower of the field."

I believe that *Papaver rhoeas* is the most appropriate selection. In the wild, poppies are common in open, well-drained areas and often have been considered weeds among farmers. They are seen in open areas in the Holy Land, and other parts of the world. In Europe, poppies have been made famous in poetry about Flanders field. Poppy is the common name for a small family of herbaceous flowering plants that grow primarily in the North Temperate zone. Poppies make up the family Papaveraceae in the order Papaverales.

There are rather different types ranging from the Oriental poppy that is widely cultivated as an ornamental with many color forms to the Opium poppy from which morphine, codeine, and heroin can be made. Because of today's drug problems, it is best to avoid cultivating that type.

The common poppy is an annual with bluish-green leaves that has hairy, fairly large leaves and several somewhat long stems that bear flowers. When the bright red flowers open, they shed the two-leaved calyx. This flower typically opens in the morning and closes in the evening, which botanists believe may be a trait linked to its habitat.

Rev. W. Wilkes discovered a poppy with a thin white edge growing among a mass of corn poppies in the 1880s, as legend says. He saved seeds and developed the Shirley poppy, named after his home town. It is silky with waved petals in shades of crimson, salmon, rose, and mixed, a delight to see. The Icelandic poppy, *Papaver alpinum,* is one excellent choice for temperate area gardens. You may also select Poppy Oriental, *Papaver orientale*, as a garden delight. The Shirley double-mix poppy, *Papaver rhoeas*, from Johnny's Seeds, offers gorgeous double blooms. The flowers are in red, pink, rose, and salmon with white accents. It prefers cool weather. Mass plantings in beds and borders make a dramatic display with plants 24 to 30 inches tall and blooms 2½ to 8 inches across. The lovely short-lived cut flowers last longer when stems are seared with a flame.

Grain is humankind's oldest, most fundamental food crop. In ancient Palestine barley and wheat were basic foods. Then, as now, wild corn poppies, *Papaver rhoeas,* were often seen as a weed in fields of grain. It usually blooms when grain is still green, before harvest time. The corn poppy, also *Papaver rhoeas,* is the common field flower of Europe and symbol of World War I, giving it the name "Flanders's poppy." It has naturalized through North America.

Long blooming, it provides cheerful color to any garden spot. This poppy prefers a moist, well-drained soil and can tolerate low fertility. It grows 12 to 24 inches tall.

Plant Profile: Poppies have appealed to humans for centuries, in locations far beyond the Holy Land. Poppy leaves are deeply divided and arranged in a rosette around the base of a short stem. Flowers have two to four sepals in outer floral whorls and twice as many petals as inner floral whorls. They grow 2 to 3 feet tall, depending on type. Other types of related poppies also deserve consideration. Appealing color photos in mail-order catalogs will allow you to pick those that please you best and fit into your garden conditions.

Growing Tips: Shirley or Flanders Field poppies are hardy annuals. That means they tolerate cold temperatures. Seeds germinate best in cool soil and should be planted in late fall or earliest spring outdoors. Mix seed with fine sand and sow thinly. Cover seeds lightly with soilless mix or sandy soil. Thin to stand 8 to 10 inches apart for best performance. Many poppies will self-sow once they have been added to your garden, so they can perpetuate themselves to some degree. They prefer fertilized, well-drained soil and sunny locations. Remove spent blooms and seed capsules weekly to extend flowering time. Oriental poppies, *P. orientale,* are perennials as are alpine and Iceland poppies. Use seed in spring or set out started or dormant plants in the fall with tops 3 inches deep. Plant container-grown ones at soil level. Poppies can spread so divide clumps in autumn when foliage has died back.

ROCKROSE—*CISTUS SP., CISTUS CRETICUS L.*

ROSE, RUGOSA—*ROSA RUGOSA*

Rose is the common name for a family of flowering plants that surprisingly includes important fruit and ornamental species. In Ecclesiasticus we read of the beauty of wisdom, which is compared with the roses of Jerico. That is one reference we share as we enjoy them. "I grew tall like a palm tree in En-gedi, and like rose plants in Jerico" (Ecclesiasticus 24:14).

Far back in ancient Chinese history, roses were grown for their beauty as well as for cosmetic purposes and perfumes. Oriental peoples have always favored perfumed flowers, judging from early writings. Roses are one of the world's favorite flowers, without any doubt. Four species of the genus Rosa are native to Palestine. Two are small alpine shrub-like plants found on Mount Hermon and Mount Sinai. *Rosa canina* is common in many temperate countries. *Rosa phoenicia* is a more restricted habitat plant found in eastern Mediterranean countries. This famed old rose has multibranched, thorny stems with large whitish flowers growing in clusters.

American native wild roses and *Rosa rugosa* are the most similar in appearance, judging from comparisons of plants and photos. I opt for either as my choice because in its meadow or roadside setting or more formal garden, a simple five-petal blossom is beautiful in its simplicity. Equally important, Phoenician roses are not available so we must seek the best prospects among look-alikes to represent roses of the Holy Land.

Frankincense and myrrh are both fragrant resins that are produced by trees or shrubs. Another resin gum, sometimes mistakenly called myrrh, is ladanum. It is secreted from the bushy native Holy Land rockrose stems and leaves during daytime heat. These fragrant droplets were collected in olden days and used to make perfumes. Scholars are generally agreed that the myrrh mentioned in Genesis 43:11 is most likely the ladanum resin obtained from rockroses, the Cistus species which grow rather abundantly in the Mediterranean region. They are evergreen bushes with showy rose-like flowers that have five crinkled reddish, pink, or white petals. Other scholars point out that true myrrh is a resin obtained from a tropical African bush.

In various translations of the Song of Solomon 2:1, "I am a rose of Sharon," debate continues whether that was meant to identify a true rose, because so many linguists, botanical, and Biblical specialists say it is more likely a crocus, narcissus, or tulip. However, it is true that the dog rose and Phoenician rose also grow in the Holy Land. Whether they are actually mentioned in Scriptures or not, it seems appropriate to include a representative rose in Biblical gardens for their appealing beauty. The rose family is in an order with 24 other families, sometimes referred to as the rose order.

Worldwide in distribution, the family actually contains many of the most important fruit trees known: apple, apricot, almond, cherry, pear, peach, plum, nectarine, and quince. The bramble bushes, raspberry, blackberry, and dewberry are related. Roses that specifically relate to Biblical times, sites, and writings also should be focused on here. Roses make up the family Rosaceae of the order Rosales. Generally the classes of old roses are based on selection from one or two ancestral species. Today there are hybrid perpetuals, polyantha roses, tea and China roses, contemporary hybrid tea roses, floribundas, and others. The complexity of rose history and culture is in a class by itself, well beyond the purpose of this book.

Yet some history of this most popular family of flowers seems in order. The rose has been grown for its fragrance and beauty since ancient times. Some are still raised in their most natural form or similar cultivars. Because of the popularity of the rose in all parts of the world, hybridizers have been busy for centuries. They combed the earth to find new roses to cross with existing ones. In the process through the centuries, more than 20,000 cultivars have resulted. They are classed as old roses or contemporary roses. Several hundred new rose cultivars are introduced each year. More than 20 million rose plants are commercially grown annually for cut flowers and more than 40 million for landscape and ornamental use.

The Phoenicia rose, *Rosa phoenicia*, is probably the one that is most deeply rooted in the Holy Land. However, availability of such a rare rose leads us to suggest other types that are similar in appearance.

Plant Profile: *Rosa canina*, or dog rose, grows 4 to 6 feet tall. It is hardy, flowers early in summer, and has many similar wild relatives with lovely single flowers. Wild roses also can be transplanted from natural sites to home gardens, if they seem most suited to you. Naturally you should ask permission of the property owner or town road department if wild roses are growing on town roadsides.

Growing Tips: Roses may be grown in any good, well-drained soil. Although different varieties respond better to certain soil and climate than others, generally sandy soils are better than clay soils. Plant in good garden soil in sunny location. Some fertilizing helps, but wild type roses are usually very hardy. Prune back to keep them from overgrowing. *R. damascene*, the Damask rose, is a hybrid between *R. phoenicia* and *R. gallica*. It bears semidouble pink blooms in early summer.

SALVIA-SAGE—*SALVIA OFFICINALIS*

Sage is the common name for the genus Salvia of the family Labiatae. The genus is a member of the mint family. Although usually considered an herb, many species are cultivated as ornamentals for their attractive flowers and foliage. Common sage is native to the Mediterranean region and is widely grown in many countries for its leaves, which contain the typical pungent sage oil used in seasonings.

> And he made the candlestick of pure gold: of beaten work made he the candlestick; his shaft, and his branch, his bowls, his knops, and his flowers, were of the same: And six branches going out of the sides thereof; three branches of the candlestick out of the one side thereof, and three branches of the candlestick out of the other side thereof (Exodus 37: 17–18).

Through the ages, artists often have borrowed from nature as they created their works of art. The Egyptians, Greeks, Romans, like the Hebrews, have adorned buildings with replicas of plants and flowers. According to Biblical scholars, this passage from Exodus can be traced to one popular herb that was common throughout the Holy Land. That plant is sage.

The Judean sage, *Salvia judaica*, grows to 3 feet tall and can be found today in most of Israel. The stems are four-angled and stiff with paired leaves. When pressed flat, sage is likened by Biblical scholars to the seven-branched candlestick that is the traditional Jewish symbol, the menorah. If you examine a sage plant with its central spike and three pairs of lateral branches, you will notice that each bends upward and inward in a symmetrical pattern. On the branches are whorls of buds, which perhaps gave the artist the knops on the Biblical golden candlestick.

A perennial of dry areas with blue flowers, sage matures 15 to 35 inches tall. Fragrant leaves were used for flavoring as are modern species of sage today. Siding with the scholars, that the sage was and is a plant of the Bible, we also know it has proved useful as a flavoring as well as for medicinal purposes for centuries.

Plant Profile: Sage is a distinctive herb with a pungent aroma. It is a hardy perennial that will grow 3 feet tall, bearing lavender to whitish flowers. The oval leaves, which may be a few inches to several inches long, are grayish green in color and somewhat coarsely textured. There are, of course, other types of sage, such as the golden sage with yellow variations on leaf edges. The garden sage is the one to grow, however, to be closest to the true Biblical plant.

Growing Tips: It is not possible to find original *S. judaica*, and collecting seeds from wild plants and bringing them into other countries is prohibited without special licenses. Alternatives are *S. pratensis* and *S. officinalis*, the common sage, a perennial, which grows 2 to 3 feet tall in rounded shape. Plant seed in pots indoors or outdoors after Memorial Day. You also can cut shoots from an established bush in spring and root in sandy soil or soil mix. Pruning back each year helps maintain attractive compact growth. Sage and lavender look similar, so consider both in your garden plantscapes. Plant several seeds about one quarter inch deep in early spring. Be patient as you water the seedbed regularly, because sage, like other herbs, may require 3 to 4 weeks to sprout. Thin seedlings 10 to 12 inches apart. Because sage begins its life slowly, keep weeds under control to eliminate any competition for moisture or nutrients.

Indoors, sage prefers a sunny window that faces south or east. Place 3 to 6 seeds in a 6-inch pot. Thin to the two strongest seedlings. Sage will reward you well without much care, once it has rooted well. In fact, overwatering can be harmful, indoors or outdoors. Mulching outdoors is helpful, but don't water unless soil is really dry. Normal rainfall is usually satisfactory.

You can harvest sage by picking the leaves periodically. To improve the quality of the leaves, prune the woody stems occasionally. This encourages new branches to form, which will provide more tender young leaves. To dry sage, cut sprays or bunches and hang them in a cool, dry spot. You can also strip the leaves from the stems and place them on a clean screen or sheet to dry in the sun.

Sage leaves are mainly employed in stuffings and sausage, as well as with meat and poultry flavoring. This pungent herb adds a piquant or sharp taste to vegetables. Many recipes call for its use with beans, in stews and soups that have their origin in Mediterranean lands. When the leaves are crisp and brittle, crush them and store them in airtight glass jars for future cooking fun and flavor.

GLOBE THISTLE—*ECHINOPS VISCOSUS DC*

Although not specifically named in the Scriptures, the globe thistle is commonly seen in the Holy Land, as it has been for centuries. Both the golden globe thistle, and other more common blue and purple ones, also grow there. In the times when our ancestors settled in the Holy Land, one judge, Gideon, led his people against attacking Midianites. In the middle of the battle, they needed food but the people of the nearby town of

Succoth refused to feed them until they could be sure of who might win that battle. Gideon threatened to punish the people by scourging them with thistle, as we see in Judges 8:7: "Well then, when the Lord has given Bebah and Zalmunna into my hand, I will flail your flesh with the thorns of the wilderness and with briers." When that battle ended, Gideon returned and may indeed have used thistles to thrash the people.

We also think of thistles as thorny problems in gardens. They certainly do have nasty thorns and sharp leaves as farmers and gardeners well know. The Syrian thistle, *Notobasis syriaca,* and the Holy thistle, *Silybum marianum,* are annuals that produce tall stems covered with small spiny leaves. These have pink or white flowers. The globe thistle *Echinops viscosus* is a perennial plant with stout, spiny stems that appears in early summer. It produces spiny heads of purplish blue to violet flowers. The golden thistle, *Scolymus maculatus,* is a widespread tall annual with a rigid, stout stem. It has leaves spread along the stem, leathery and divided into spiny lobes. The composed heads are yellow or golden color. In the New Testament, the thorns that grow in grain fields, as noted in Matthew 13:7, might indeed be golden thistles. It is a noxious weed in fallow fields and meadows in the lower altitudes of the Holy Land. Other references to thistles can be seen in Job 31:39–40: "If I have eaten its yield without payment and caused the death of its owners; let thorns grow instead of wheat, and foul weeds instead of barley."

More references can be identified in other Scriptures too. In Isaiah 34:13, we read: "Thorns shall grow over its strongholds, nettles and thistles in its fortresses." Jeremiah 4:3 warned men not to sow "among the thorns" but in broken fallow ground. Actually, thistles are common around the world. Thistles belong to the family Compositae. Another species of globe thistle is *Echinops sphaerocephalus.* The common or bull thistle is *Cirsium vulgare,* and the Canadian thistle is *Cirsium arvense.* The term "thistle" is applied in a restricted sense to plants in several genera that possess spiny leaves and branches and sharp, spiny bracts around the flowers. The common or bull thistle has purple flowers. The Canada thistle has smaller lilac or white flowers.

The musk thistle that is common in meadows in northeastern United States has purple flowers. The classic globe thistle of the Holy Land has bright yellow blooms.

The star thistle, *Centaurea calcitrapa*, is one of the thistles that grows throughout the Mediterranean area. As we note in Genesis 3:17–18, God curses the ground and says it will bring forth thistles and thorns as punishment for the first man and woman breaking the rule about not eating from the Tree of Life. As we think about thistles, perhaps the star thistle should be included with its deep purple flowers, or as close a representative in appearance as we can conveniently find. Other botanical scholars focus on the so-called Holy thistle, *Silybum marianum*, also called lady's thistle or milk thistle. Some say this is the thistle talked about in Genesis 3:17–18 and also in Isaiah 34:123. It is no doubt an attractive and ornamental plant with typical thistle flowers. Remember if you grow it, be wary. It is a thistle, and can spread out and snag clothes as well as scratch hands and arms. A small blue globe thistle, *Echinops banaticus,* has intense blue, spiky round flowerheads produced in profusion on the ends of branched stems. This is an excellent cut flower and dries well for winter bouquets. It blooms in summer, prefers well-drained soil, and will self-sow.

Although not a thistle, the Armenia basketflower, *Cengtaurea macroephala*, has the appearance of a yellow globe thistle. An impressive plant that captures attention, it is a native of the Caucasus and grows 40 to 48 inches tall. A fascinating look-alike to the yellow globe thistle, but without prickly leaves.

Plant Profile: Thorny, rugged-looking, and dramatic plants, mostly with purple flowers, are represented by several types; all thorny pests according to farmers through the centuries.

Most thistles grow 2 to 4 feet tall, with deeply cut, prickly leaves. Distinctive flowers are spherical, from golf ball size to larger. Most bloom in midsummer to fall. Plants may be *E. exalatus, E. humilis, E. ritro,* but they all have similar appearances to represent thistles of the Scriptures, if you really want thorny plants. Happily, goldfinches love the thistle seed each fall, which somewhat makes up for the plants prickly problem.

Growing Tips: Although thistles may trace their roots to the Mediterranean, today various types grow globally. Seeds can be sown in ordinary soil in early spring. Thin out to stand 18 to 24 inches apart because bushes mature large. All types prefer sunny locations.

Established plants tolerate dry periods, but younger ones like moisture to get started well.

Flowers are useful in dried arrangements. Cut before they open and dry upside down or in silica gel. Clumps will last for years once they set a firm roothold. Remember that thistles are thorny, so plant them away from walkways, and wear gloves when pruning. Veitch's Blue is a dramatic blue globe thistle that grows 2 to 3 feet tall. Plant 1 foot apart in Zones 3 to 9 in which most types of thistles thrive.

VIOLET—*VIOLA ODORATA*

Violets have significance to all of us, according to Rev. Marsh Hudson-Knapp, founder of the Biblical Garden at Fair Haven Congregational Church in Vermont. "The first letter of Peter urges us to deeply root ourselves in God and God's word," he notes. "In time, 'The grass withers and the flower falls, but the Word of the Lord abides forever,' as we learn from 1 Peter 1:24–25. The fragile blossoms of the violet remind us of our own vulnerability in contrast with the eternal nature of God who sustains us," Rev. Hudson-Knapp believes.

Violets thrive around the world. You can find them in shades of blue, violet, purple as well as white and yellow. Among wild violets, some like sun while others prefer shade. Common violets are found from Maine to Georgia in moist meadows and stream sides. Transplant violets by dividing clumps in spring or fall. Most root readily and spread themselves by seeds through the years. One of our early beds now is covering the ground beneath a red barberry bush, providing spring bloom and crowding out weeds with its heart-shaped leaves during the summer.

An alternative for your garden could be the choice of violas or pansies as symbolic of violets in your garden. Johnny-jump-

ups are another potential. Violas prefer cooler weather to perform best, but will provide blooms throughout most of the
summer and into fall. In mild winter areas they also may over-
winter. With mulching, ours have done so and begin blooming
along with daffodils in the spring. They also perform well in
window boxes and planters, especially to add multicolors for
deck and patio display. One new hybrid called Yesterday, Today
and Tomorrow actually changes color. Yesterday they were almost white, today they are light blue, and tomorrow will become deep blue. Each flower goes through three changes of
color, so these are an extra delight for beds or containers.

Plant Profile: Violas or pansies remain among America's favorite flowers. Violas and pansies have been garden favorites for
centuries, and were known by names such as Call-Me-To-You
and Faces-in-a-Hood. They are cool weather plants that grow 6
to 8 inches tall, have compact growth habit, and are now available in a wide range of colors.

Growing Tips: Sow seed indoors 6 to 8 weeks before setting
plants outdoors in northern areas. In the South, sow in the fall
for winter and spring bloom. They prefer cool weather.
Removing spent blooms prolongs flowering. If foliage gets leggy
in warm weather, simply cut back to 2 inches of the base to force
new flowering shoots. Ideal for combining with spring bulbs to
carry color into summer and fall.

WATER LILY—*NYMPHAEA LOTUS* (WHITE), *NYMPHAEA CAERULEA* (BLUE)

Water lilies provided a source of great beauty in ancient times.
King Solomon used their shapes to adorn the tops of columns in
God's holy temple in Jerusalem as we know from I Kings 7:19. It
is a fact that water lilies grew in Egypt and Palestine in ancient
times as they do today. Fact is, three varieties of water lilies existed in the Holy Land. They included the common water lilies,
Nymphaea alba and *Nymphaea caerulea*.

Obviously, you would need a pond or similar water growing
environment to grow water lilies. They do require special care,

even if you have a small pond on your property. Because these water lilies are tender to cold and can't take freezing, you may wish to substitute native wild water lilies, or newer, hardier cultivated types. Many are available today. Water gardens are among the fastest-growing new gardening projects across America this decade, according to recent reports. The soothing sounds of trickling water are being used in restaurants and office buildings as natural backgrounds that seem to please customers and office workers alike. For generations the Japanese have created exotic water gardens as an important part of their home horticultural environments.

Rev. Hudson-Knapp and his wife designed and built a water garden as part of the Biblical gardens at their church. "People walked past our gardens every day, but when we installed our Biblical water garden, walkers stopped day after day to find peace watching the waterfall, to look for the fish and frog, and consequently to talk with other adults and children who were also stopping. Our gardens shifted from being something pretty to a gathering place for building community."

Some gardeners prefer to grow as close to Biblical lore as they can. You can find details about water lilies from the key source in my favorite garden websites. From Lilypons Nursery you can obtain lilies and all the necessities for creating your water lily environment. You might then also expand your view to the reeds and rushes and cattails too, as some church gardens have done to commemorate similar scenes in the Holy Land.

Plant Profile: Water lilies were common in ancient Egypt, where they grew along the Nile. Two wild species include *Nymphaea lotus* and *N. caerulea*. The first is white and the smaller caerulea has blue flowers. Both are fragrant. Subtropical water lilies must be grown in a warm water area 1 to 2 feet deep where rhizomes can be submerged in tubs of warm, rich humus content soil. Because this is impractical for most gardeners in the United States, I recommend you select similar white-flowered or colored species if you prefer as representative of this meaningful plant. Many dramatic hybrids have been developed.

Growing Tips: Some gardeners report success starting water lilies from seeds or transplanting from the wild. Best bet is to obtain rhizomes from water garden specialists like Lilypons, where you have a wide selection of top-quality plants.

FLOWERS IN KEY BIBLICAL GARDENS

For reference, here are lists of the flowers from among all plants in various major Biblical gardens. You'll also find the suggested Scriptural quotations as provided by those who created and tend these special gardens. Naturally, all gardens are in various stages of growth, with some plants reaching maturity, others being added, and a few eliminated at times as gardens are re-designed.

Flower	Latin Name They Use	Their Scriptural Reference
Almond	Amygdalus communis	Numbers 17:8
Anemone	Anemone coronaria	Matthew 6:28
Apricot	Prunis Armeniaca	Genesis 2:17
Buttercup	Ranunculus asiaticus	Psalms 4:5
Saffron Crocus	Crocus sativus	Song of Solomon 4:13
Cyclamen	Cyclamen persicum	Song of Songs
Daffodil	Narcissus taxetta	Isaiah 35:1
Daisy, Shasta	Bellio silvestris cyr	Psalm 45
Dandelion	Taraxacum officinale	Exodus 12:8
Dove's Dung	Ornithogalum umbellatum	II Kings 6:25
Hibiscus	Hibiscus syriacus	Job 30:4
Hollyhock	Alcea setosa	Matthew 6:38
Hyacinth	Hyacinthus orientalis	Song of Solomon
Iris	Iris Palestina	Song of Solomon
Iris, Yellow Flag	Iris pseudacorus	Hosea 14:5
Jonquil	Narcissus jonquilla	Song of Solomon 4:5
Larkspur	Delphinium ajacis	Song of Solomon 6:24
Lavender	Lavandula stoechas	Song of Solomon 6:24
Water Lily	Nymphaea caerulea	I Kings 7:22
Linum—Flax	Linum perenne	Proverbs 31:13
Loosestrife	Lythrum salicaria	Song of Solomon1:14
Lupine	Lupinus augustifolius	Song of Solomon
Mallow	Althea officinalis	Song of Solomon
Myrtle	Myrtuscommunis	Nehemiah 8:14-15
Narcissus	Narcissus tazetta	Song of Solomon
Poppy	Papaver orientalis	Jeremiah 8:14
Rose	Damescena	Isaiah 35:1
Rose, Phoenician	Rosa Phoenicia	I Ecclesiasticus 24:14
Rose of Sharon	Tulipa sharonensis	Song of Solomon
Sternbergia	Sternbergia lutea	Song of Solomon
Globe Thistle	Echinops viscosus DC	Judges 8:16
Tulip	Tulip montana, T. sharonensis	Song of Solomon 2:2
Violet	Viola odorata	Song of Songs

Flowers of the Bible, St. Gregory's Episcopal Church, Long Beach, California

Flower	Latin Name They Use	Their Scriptural Reference
Myrtle	*Myrtus comminus compacta*	Nehemiah 8:15
Flax	*Linum usitatissiumum*	Proverbs 31:13, John 19:40
Dandelion	*Taraxacum officinale*	Exodus 12:8
Tulip	*Tulip montana*	Song of Solomon 2:12
Cyclamen	*Cyclamen persicum*	Luke 12:27
Narcissus	*Narcissus tazetta*	Song of Solomon 1:2, Isaiah 35:1
Anemone	*Anemone coronaria*	Matthew 6:28–30
Poppy	*Papaver rhoeas*	Isaiah 40:6-8, I Peter 1:24–25
Sternbergia	*Sternbergia lutea*	Isaiah 40: 6–8
Yellow Flag	*Iris pseudacorus*	Hosea 14:5
Sweet Flag	*Acorus calamus*	I Kings 10:10, Ez. 17:19
Sea daffodil	*Pancratium maritmum*	Ecclesiasticus 50:8
Apricot	*Prunus armeniaca*	Genesis 31:7, Proverbs 25:11
Crocus	*Crocus sativus*	Isaiah 35:1–2
Iris	*Iris persicum*	Isaiah 60: 6–8
Star-of-Bethlehem	*Ornithogalum umbellatum*	II Kings 6:25
Sage	*Salvia judaica*	Exodus 37:17–23
Mustard	*Brassica nigra*	Mark 4:30–32
Turk's Cap Lily	*Lilium chalcedonicum*	Song of Solomon 5:13
Rockrose	*Cistus incannis*	Genesis 43:11

Prince of Peace Lutheran Church, Augusta, Maine

Flower	Latin Name They Use	Their Scriptural Reference
Anemone	*Anemone coronaria*	Luke 12:27
Crocus	*Crocus sativus, Crocus purpureus*	Isaiah 35:1
Crown Daisy	*Chrysanthemum coronarium*	I Peter 1:24–25
Cyclamen	*Cyclamen persicum*	Matthew 6:28
Mallow	*Malva alcea, Malva sylvestris*	Job 6:6–7

Flower	*Latin Name They Use*	*Their Scriptural Reference*
Flax	*Linum usitatissimum*	John 19:40, Genesis 41:42
	Linum humile	II Chronicles 1:16, others
Iris	*Iris pseudacorus*	Exodus 30:23, Song 4:14
Lily	*Lilium candidum*	Matthew 6:28
Mint	*Mentha piperita*	Matthew 23:23
	Mentha spicata	Same
Mustard	*Brassica nigra*	Mark 4:30-32
Myrtle	*Myrtus communis*	Leviticus 23:43
Narcissus	*Narcissus tazetta*	Song of Solomon 2:1–2
Poppy	*Papaver rhoeas*	Isaiah 40:6–8, I Peter 1:24,
		John 19:29–30, Matthew 27:34
Sage	*Salvia judaica*	Exodus 37:17–18
Tulip	*Tulip montana*	Song of Solomon 2:1–2
	Tulip greigii	Same
Rose	*Rosa phoenicia, Rose chinensis*	Song of Solomon 2:1–2
Yarrow	*Achillea santolina*	No verses

Cathedral of St. John the Divine, New York

Flower	*Latin Name They Use*
Anemone	*Anemone blanda*
Crocus	*Crocus sativus*
Flax	*Linum perenne*
Hollyhock	*Althaea rosea*
Hyacinth	*Hyancinthus orientalis*
Iris, Yellow Flag	*Irus pseudacorus*
Lily	*Lilium regale*
Mallow`	*Malva moschata*
Mint	*Menta longifolia*
Narcissus	*Narcissus tazetta*
Poppy	*Papaver rhoeas*
Rockrose	*Cistus sp.*
Rose, Dog	*Rosa canina*
Rose, French	*Rosa gallica cv trigintapetala*
Sage	*Salvia officinalis*
Tulip	*Tulipa fosteriana and greigii*

Flower	Represents	Their Scriptural Reference
Anemone	Lily of the Fields	Matthew 6:28–30
Crocus, fall	Saffron	Song of Solomon 6:4
Day lily	Lily of the Fields	Matthew 6:28-30
Flax	Linen, flax	Exodus 9:31
Iris	Lily	Hosea 14:5
Mint	Biblical Mint	Luke II
Madonna Lily	Lilies	Song of Solomon 5:13
Mustard	Mustard	Matthew 13:31–32
Myrtle	Myrtle	Nehemiah 8:15
Narcissus, tazetta	Rose	Isaiah 35:1
Nasturtium	Bitter herbs	Numbers 9
Rose de Provence	Rose	Isaiah 35:1
Rose, Climbing	Rose	Isaiah, 35.1
Sage	Candlestick	Exodus 37
Star-of-Bethlehem	Dove's Dung	II Kings 6:25
Thistle, Globe	Thistles	Judges 8:7

Neot Kedumim, Lod, Israel

English Name	Latin Name	Biblical/Talmudic Identification
Almond	*Amygdalus communis.*	*shaked, luz*
Buttercup, Red	*Ranunculus asiaticus*	*nitzan, nurit*
Cattail	*Typha domingensis*	*soof*
Cyclamen	*Cyclamen persicum*	*rakefet*
Hawthorn	*Crataegus aronia*	*tapuakh/uzrad*
Iris, Purple	*Iris astropurpurea iris*	*argaman*
Lavender, Blue	*Lavandula stoechas*	*ezov kokhli*
Lily, Madonna	*Lilium candidum*	*shoshan*
Lily, Sand	*Pancratium maritimum*	*khavatzlet hakhof*
Loosestrife	*Lythrum salicaria*	*shanit*
Mint	*Mentha sp. koranit*	*hanakhal*
Narcissus	*Narcissus tazetta*	*shoshanat ha'amkim*
Rockrose	*Cistus sp.*	*lot*
Rose, Dog	*Rosa canina*	*shoshana vered*

English Name	Latin Name	Biblical/Talmudic Identification
Sage, Jerusalem	Salvia hierosolymitana	Boiss moriah
Sage, Judean	Salvia, Judaica	moriah
Squill	Urginea maritima	khatzav
Thistle, Golden	Scolymus maculatus	khoakh
Tulip, Sharon	Tulipa sharonensis	hasharon
Yarrow	Achillea fragrantissima	samtarin

First Congregational Church of Fair Haven, Vermont

Flower	Latin Name They Use	Their Scriptural Reference
Almond	Prunus dulcis	Numbers 17–7
Almond, Flowering	Prunus triloba	Ecclesastes 12:5
Anenome, Crown	Anemone coronaria	Matthew 6:28
Apricot	Prunus armeniaca var.	Genesis 2:9
Cattail, Dwarf	Typha minima	Isaiah 19:6
Crocus, Saffron	Crocus sativus	Isaiah 35:1
Daffodil, Trumpet	Narcissus sp.	Matthew 6:29
Daisy, Shasta	Chrysanthemum sp.	I Kings 6:29
Dandelion	Taraxacum officinale	Exodus 12:8
Flax, Blue Perennial	Linum perenne	John 19:40
Hibiscus	Hibiscus syriacus	James 1:9–10
Hollyhock	Alcea or Althaea setosa	Job 6:6–7
Hyacinth	Hyacinthus orientalis	Song of Solomon 6:2–3
Hyacinth, Grape	Muscari sp.	Song of Solomon 6:3
Iris, Blue Flag	Iris versicolor	Hosea 14:5
Iris, Purple	Iris astropurpurea	Hosea 14:5
Iris, Varigated	Iris sp.	Hosea 14:5
Iris, Yellow Flag	Iris pseudacorus	Hosea 14:5
Larkspur, Delphinium	Delphinium cultorum	Luke 12:27
Lily, Madonna	Lilium candidum	Song of Songs 2:1–2
Lily, Water	Nymphaea lotus/caerulea	I Kings 7:19
Lupine	Lupinus	James 1:9–10
Mallow	Malva moschata	Job 6:6
Mint	Menthe sp.various	Luke 11:42
Cornflower	Centaurea montana	Genesis 3:17–18

Flower	Latin Name They Use	Their Scriptural Reference
Mum, Hardy	*Chrysanthemum sp.*	James 1:9–10
Mustard	*Brassica nigra*	Matthew 17:20
Narcissus	*Narcissus tazetta*	Matthew 6:30
Onion	*Allium cepa*	Numbers 11:5
Poppy, Shirley	*Papaver rhoeas*	Isaiah 40:6
Poppy, Icelandic	*Papaver alpinum*	Isaiah 40:8
Poppy, Oriental	*Papaver orientale*	Isaiah 40:6
Red Buttercup	*Ranunculus asiaticus*	1 Peter 1:24
Rose, Rugosa	*Rosa rugosa*	Wisdom 24:14
Sage, Dwarf	*Salvia officinalis*	Exodus 37:17–18
Star-of-Bethlehem	*Ornithogalum umbellatum*	II Kings 6:25
Thistle, Globe	*Echinops viscosus DC*	Judges 8:7
Thistle, Spanish Globe	*Centaurea macrocaphala*	Matthew 7:15
Tulip, Red	*Tulipa sp.*	Song of Solomon 2:12
Violet	*Viola odorata*	James 1:10

Chapter Four

Pick the Right Site and Prepare the Growing Ground Properly

The most basic first step for gardening success is to pick the right location. That sounds logical. It is also easy to say except when you are limited by available space, poor exposure to the sun, and have problems with nearby trees, buildings, or other growing situations.

Actually, with at least six hours of sun each day, access to water to give thirsty plants their needed drinks each week, and room to work the area, even in poor soil you can make blooming miracles happen. We've seen the unlikeliest areas bloom gloriously when a person sets his mind to good gardening. Surprising success can be achieved.

Examine your garden areas at different times of day to see how the sun flows over it. Where possible, pick a sunny, well-drained location with at least six to eight hours of direct sun each day. Southern exposures are best. Experienced gardeners know southern exposures provide maximum sun for peak plant growth. If you don't have a southern exposure, next best in order of desirability are eastern, then western, and finally northern exposure plots because they provide less good sunlight.

AVOID PROBLEM AREAS

Avoid areas where shadows from trees or buildings block the sun too long each day. In areas near tall trees or paved area runoff, you'll need to make cultural adjustments, such as directing excess water and street pollution wastes away from the garden areas and adding extra fertilizer for your flowers. Also, avoid nearby hedges or trees with shallow roots that draw water and nutrients from the soil.

Keep beds running north and south when possible so that each row has an equal amount of sunlight. If you have slopes, consider contour planting as farmers do, to avoid soil washing away in storms. Rows up and down the slope would easily erode valuable topsoil during heavy rains. If you do naturalized flowers mixed with shrub and tree plantings in your landscape or otherwise grow near tree roots, plan to fertilize periodically to nourish your flowers properly.

THINK SWEET SMELLS

Remember that it is best to plant fragrant flowers upwind so that their sweet scents drift to the areas where you'll be sitting, entertaining, or otherwise enjoying your Biblical flowers. To determine usual wind direction, wet your finger and hold it in the air when the wind blows. The cooler side is where the wind comes from. Or observe the direction tree branches blow.

When you take care of your soil, it will take care of you and your garden plants. That's a truth that goes back to Babylonia and beyond. The basic principle for productive gardening is improvement of the soil. The better the humus and soil conditions you can build, the better your plants will grow. Healthy soil lets you grow healthy plants.

There is a key factor that stands out in this effort: Insects dislike healthy plants! They actually prefer sickly and weak plants. When you improve the good earth and fertilize your plants, you'll have strong, healthy growth, plentiful blooms, and few insects.

LEARN ABOUT YOUR SOIL

An understanding of soils and how to improve them is basic to improving growing conditions, wherever you live and garden. No matter what you have, from clay to sandy soils, from construction debris and fill around a new development home to wet spots in your land, soil can be improved. Of course, nothing happens overnight. But you can make immediate and useful improvements this year and continue on that natural soil-building process month after month, year after year. The results will astound you.

Soil organic matter is the key. It is dynamic. It is alive. Soil changes and improves through continuing decomposition of organic materials, in it and especially as you add new raw materials, which fortunately are abundantly available.

Soil comes in various types and quality. You shouldn't let the color of soil fool you. Rich-looking dark, black soil can be low in nutrients. Reddish, sandy soils can be high in them. No matter where you live or what soil you find there, you can improve it, rebuild it, and upgrade your growing conditions immensely.

All soils have several things in common. They contain organic matter, water, air, and minerals. The proportion of these elements vary, but these components remain essentially the same. Bacteria and fungi, which also occur in your living soil, are vital contributors to soil formation. They live on animal and plant residues. They break down complex compounds into simple forms. Nitrogen-fixing bacteria, for example, in the nodules of legumes actually take nitrogen from the atmosphere and help make it available in the soil for future plants. Planting legumes such as clover and alfalfa as rotation cover crops and then plowing or tilling them under helps add nitrogen to garden soil.

SOIL IMPROVEMENT IS EASY

All soils can be improved with the proper treatment. Our initial objective is to aim for a balance in texture and porosity. Texture refers to the size of the majority of particles making up the soil and ranges from microscopic clay particles up to small stones and gravel. Porosity refers to the pores in the soil though which water, nutrients, roots, and air most move. Balance is another useful term. Good gardening is a matter of balance. Excesses are what throw gardens off balance and us out of balance in life. We should avoid too much of anything.

Pick up a handful of rich, warm soil in spring. Crumble it in your hands. If it crumbles freely in your palm, you are approaching the ideal texture. Naturally, there are unseen factors, such as nutrient levels in the soil. But the consistency of the growing medium is of underlying importance. The closer you have or can rebuild soil to a granular feel, with clusters of soil that easily shake apart, the better your garden will grow.

Too often, gardeners are tempted to purchase "topsoil" or "loam" as a quick-fix approach to lawn and garden building. Rich-looking topsoil, or so-called loam, may be offered at low prices by some contractors. Some obtain this good-looking, dark soil from dredging silt in rivers and ponds and mixing it with sand. Be wary, you may be buying problems with harmful residues found in river bottoms.

Reputable nurserymen, garden centers, and similar suppliers are also aware of problem soil in your area. One example might be residues of pesticides or herbicides in farm fields from which topsoil is being sold to homeowners. Ask for their recommendations for improving your growing ground. They usually want your business for seeds, plants, and trees and will answer questions honestly, provided you know the questions to ask.

Key Questions to Ask

1. Is this topsoil from old farmland?
2. Does this soil have any herbicide or pesticide residues in it?
3. Has this soil been mixed from river bottom soil?
4. Has this topsoil been mixed with animal manures to improve its fertility?

The riches held within the soil must be unlocked in order to produce a productive, bountiful garden. Compost is the result of combining organic matter and manure that decays into a valuable soil additive. It is the key that turns soil into productive land that will yield bountiful, magnificent flowers and hardy, attractive shrubs and trees. Veteran organic gardeners have long recognized the value of compost to improve soil texture and also add small amounts of nutrients to the soil too.

COMPOST MAKING TIPS

You can make compost easily and in a short time with a little extra effort. Using readily available materials, you can turn organic matter into usable compost in as few as 14 days. Soil—like the plants that grow in it—is alive! Millions of bacteria, fungi, minute animals, and other microorganisms inhabit it. There is a close relationship between the amount of life in the soil and the soil's basic fertility.

This tiny animal life reduces complex organic substances such as sugars and proteins to simpler chemical forms, i.e., carbon dioxide, nitrate, and water. In this way nutrients are made available for plants again. Your objective is to increase the life, and consequently the productivity, of the soil. Successful organic gardening is directly related to the use of compost for enrichment of the soil. The better the compost, the better the soil. As good gardeners realize, just as we need food to grow, our plants also need water plus nutrients for growth. Improving soil texture enables plant roots to move more easily through the soil to pick up the nutrients they need to grow well.

Compost can be made in many ways, and almost any type of organic material can be utilized effectively. Production of compost depends upon the decay or decomposition rate of the various materials used. Bins or pits, in the ground or above the ground or piled along garden rows, plus many other variations have been successfully utilized to hold the material.

You can pick the method that best suits your needs, i.e., the time you have available and the efforts that you want to make.

Any soil can be improved with compost, even the hard, caked subsoil often found around new home developments. The addition of humus, which is generally defined as well-rotted compost, provides the best and least expensive way to improve your soil.

Yes, you can create topsoil. One nice factor is that soil can never get enough organic matter. Good topsoil contains from 2 to 8 percent organic matter by dry weight, which is the residue of growing plant materials. You should continue to add organic matter every year in order to keep building your soil bank. Each deposit you make will yield dividends. In general, the addition of 1 inch of organic matter of humus from compost per 4 inches of topsoil constitutes a good average application.

Avoid using diseased plants for compost. Despite the fact that natural heating in the center of the pile during decomposition of organic matter may and usually does kill most disease organisms, there are still some risks. For example, don't use peony tops if they are infected with botrytis blight or iris leaves if iris borers are a problem.

COMPOST MAKING POINTERS

Ideally, compost in the process of becoming that rich, dark humus you want will not have a seriously objectionable smell. However, in any decomposition process of organic matter there is an odor. Another consideration is appearance. Compost piles do not exactly have eye appeal. If you live in an area where homes are close together, then you must consider the appearance of your compost area. Select an out-of-the-way location. You probably will want to screen the area from view with a hedge, shrubs, or perhaps a fence around it. We use a berry hedge where possible. The bushes form thick bramble patches, hiding the compost piles and providing tasty berries in the bargain.

Choose a site convenient to water so you can periodically wet down the pile or pit to keep it moist. A shady area is good, but avoid low areas in which rain collects and the ground remains soggy. While bacteria that help make compost into good humus require ample moisture, they must have air and oxygen as well. That's why turning a compost pile is helpful. It lets bacteria obtain oxygen to do their work better and faster, breaking down organic material.

The second step in preparing compost, once you have selected a site or two, is construction of the pit or pile. Remember that air circulation is important. The more air that circulates through the pile and around it, the faster the decomposition. Turning a compost pile is hard work. Good exercise, but work! If you have a front end bucket on your garden tractor, use it to turn your compost pile. Most of us do not have one so we must do hand spading or pitchfork turning to speed up decomposition by stimulating bacterial activity. It is worthwhile to have at least two piles of compost. You can use finished compost from one while still making more humus in the other pile. Finished compost, well-rotted organic material is called humus and has a crumbly feel. It is a basic soil improvement ingredient for all gardens.

TYPES OF COMPOST PILES

If you want quick compost, you must turn it by hand, using a spading form or shovel to ensure bacteria get adequate air to do their job helping break down organic materials. You also can obtain home compost tumbler bins. If you will be satisfied to develop the compost over a period of several months, you can use alternate systems. There is no single way. The best way is that method that best suits your needs and garden requirements.

Almost any organic material can be composted with the simple Indore method. This method was developed and practiced in England by Sir Albert Howard, father of the modern organic movement. It is the best, most widely used method by small and large gardeners and it is both practical and efficient.

In a nutshell, the Indore method is the simple layering of various materials. You might begin with a 6-inch layer of green organic material. Do not let the green nomenclature fool you. Grass clippings or dried leaves may be brown but they are considered "green" matter in the terminology of organic gardening.

Over this first layer of green material—clippings, leaves, old weeds, and vegetation pulled from the garden—add a 2-inch layer of manure. The objective is to add nitrogen to hasten the decay process. Cow, horse, sheep and poultry manure will do the job.

With manures considered "hot" that have a tendency to damage plants when applied too fresh without composting, including poultry and pig manure, use slightly less than you would with cow, horse, or sheep manures. It is best if the manure also includes straw, shavings, or other usual bedding materials used on most farms.

If you do not have manure readily available when you are building your compost pile, you may substitute other nitrogen sources: blood meal, digested sewage sludge, cottonseed meal, or soybean meal. Your local garden center authorities can provide details about those ingredients which are basic organic gardener materials. Most garden centers and hardware stores or chain stores with garden centers also have commercially prepared, dried manures and composts prepackaged, which are excellent to use in compost making.

The next layer in your compost pile should consist of an inch or two of garden soil taken from your garden and evenly spread to ensure effective interaction. It will have natural and useful soil bacteria in it to help with the composting process. Following the layer of soil, add a 1-inch layer of mixed rock phosphate and limestone, which you can obtain from local garden centers and even from today's mass market chain store garden departments. Top this off with manure or a soil covering a few inches thick.

You can build a small pile, just 4 feet wide, 4 feet long, and 4 feet high, or you can increase the dimensions if more material is available. As previously recommended, it is desirable to have several smaller piles so they will be in varying stages of decomposition and being finished for use in your garden.

KEEP COMPOST PILES MOIST

As you apply the layers, sprinkle them with water. All organic material going into the pile should be moist, especially if dry leaves, grass clippings, and other dry materials are being used. When you have finished building the pile, leave a depression on the top of it so that rainwater can be caught and allowed to trickle down through the layers. This helps keep them moist and you'll only need to apply a few waterings in dry periods. However, if you are in a drought period, be sure to water the compost pile once or twice a week.

Two types of bacteria will be at work in your compost. These are called aerobic and anaerobic. The first type, aerobic, needs air circulation in order to do its job. The second type, anaerobic, works more slowly and proceeds without much aeration of the pile. Turning the pile by fork or spade or otherwise providing improved aeration will quicken the entire process. Spading means work, but only two or three turnings are required before the humus is ready.

An alternative to manual aeration is to build the pile around several pipes with holes through them. This method lets the air penetrate, helps the heat build up as decomposition occurs, and saves your back from spading chores. There are many other ways to accomplish similar results; as many, in fact, as there are ingenious gardeners.

COMPOST BINS AND PITS

The elevation of composting material off the ground, thereby allowing freer air circulation beneath, provides one of the easiest ways to speed up decomposition. This can be done in several ways. Cinder blocks are handy, inexpensive, and reusable. Also, they do not rot. Space them around the area on which you will pile the organic material; one block, one space, one block, etc. On this foundation place a layer of plastic-coated fencing. By spacing cinder blocks, air can move under the wire that holds the compost. This system also lets excess rain drain out.

You can then put a wire fence around the pile or use old boards or even rows of cinder blocks to contain the organic material. Be sure to leave spaces for that vital air circulation. An area 5 by 5 by 10 feet will accommodate one large pile or two smaller ones nicely. Construction materials at hand are just as good. Field stones removed from the garden, old bricks from a dump, wire from a poultry run, or even old logs can be utilized.

One word of caution. Do not use treated railroad ties or attempt to treat the wood with chemical preservatives. Contamination from unwanted chemicals could occur. After you have expanded your efforts to garden organically and well, why risk outside pollution?

When you build a pile, with the Indore method or variations on that theme, you can make it as high as is practical. A 4- to 5-foot-high pile is probably best. That size enables you have to have several series of layers: green matter, manure, lime, rock phosphate, soil, in one easily manageable pile. If you turn the pile once a week, you can expect finished compost within two to three months, depending on weather and other variables. Several factors influence the rate of decomposition. The first is the amount of manure and nitrogen sources, plus moisture, that was included as the pile was built. Next is the amount of aeration that is provided, either through hand turning or mechanically.

A third factor is the addition of earthworms. Their tunneling and digestion of organic material hastens the process, but they should not be added until the first heating-up period has ended. That is the time when organic matter builds up heat as it begins decomposing. As this first phase of decay proceeds over a few weeks, the internal heat of the pile may reach 150 degrees F or more. The earthworms can be added after this period, when you can reach into the pile and determine that it has cooled down. I find that occurs after I've turned the pile about three times.

Bacteria, of course, are the microscopic helpers in any compost-building activity. It is beneficial to use them to inoculate the pile as it is built. Soil bacteria cultures may be purchased from various sources including Plants Alive company, a major natural gardening product firm, but the easiest way to meet this need is to save the remnants of a previous compost pile. Add the material not quite completely decomposed to the new pile or to the soil or manure layers. The bacteria will thrive and multiply and go right to work with their millions of beneficial bacterial neighbors.

Naturally spring and summer are the best times for making compost. Sun and rain, warm weather and water speed the process. Very little action takes place from November until April, except in southern areas of the country. Therefore, when a compost pile is prepared in the fall, from old leaves and debris, not much action can be expected until spring sets life in motion again.

Today, commercially available compost bins and turners are also readily available. With easy hand or mechanical cranking and turning systems, they help you achieve finished compost in a shorter period with less work.

FIELD COMPOSTING

Field composting is easier to do but takes much longer to produce the finished compost. Basically it is a layering method that needs no pit or special pile. As you gather green matter, apply it in layers to a pile or along garden rows and just let it rot by itself. Keep adding more organic material, raked leaves, grass clippings, and weeds until you have a satisfactory 4- to 6-inch layer. Then add manure or other nitrogen-containing materials from your garden center. Sprinkle on an inch or so of soil, add lime and rock phosphate, and then just let it stand. Nature will take its course and the material will slowly decompose.

Add on more green matter until there are several sequences of layers. By the following year, providing you take the time to turn the field pile at least once, you will have good humus. Even if you do not turn the pile, you can expect this field composting method to yield good humus in time. Actually, I like this system because all I must do is spread organic materials along garden rows and let them rot down in place as a natural mulch that also helps stop weed growth.

Composting is a continuous process; the remnants you apply to the newer piles help "inoculate" them with more bacteria to keep the microorganism level in high gear. Just about any type of vegetation and organic matter can be put to use to make compost. You can turn garbage into humus with little effort, vegetable parings, tops of carrots, radishes, beets, corn husks, pea pods, lettuce and cabbage outer leaves, in fact any type of household vegetation that is organic.

However, do not use kitchen wastes such as animal fats and bones. They decompose slowly and also attract dogs and other animals.

STARTING YOUR COMPOST PILE

In a given month the average family might throw away the makings of a bushel of compost. Why waste it? Especially on small plots of land, leaves and grass clippings are scarce. That's even truer in towns and cities. If you want to go organic, recycle the organic matter available to you. In organic gardening, "garbage is golden." For a starter, simply put the materials in a paper bag or wrap them in old newspaper. Then add them to your compost pile with a shovel of soil or compost over them to start them decomposing.

Many towns now prohibit the burning of leaves and brush and that means a bonus for you. Some towns collect the leaves and dump them at a sanitary landfill or city dump. You can, therefore, collect leaves from your own home town. Let the neighbors pile their leaves in your compost heap or even get the extra supplies you need for compost making from the weekly town leaf pickup. Every fall I watch for people raking their leaves, stop in, say hello, and offer to truck them away in my pickup or the trunk of our car. That adds abundant supplies to our compost piles and also provides plentiful mulching material for our beds and borders.

Leaves of all types can be added to the compost pile: Composted leaves from a single large oak or maple tree are worth as much as $15 in rich humus. That is how much that amount of humus would cost if bought as a bag of humus at a garden center. All weeds, extra leaves and stalks from your vegetables, grass clippings, hay collected from highway mowing, or bales of old, spoiled hay from farms make excellent source material. If you use a large quantity of weeds, you need to add more manure or high nitrogen sources in order to help that organic material decompose faster. Peat moss can be added to the compost pile if you wish. However, peat is best utilized right on the garden as mulch and soil conditioner.

In all organic gardening activities, keep clearly in mind that you are aiming for balance of organic materials and balance with nature. The more variety you put into the compost, the more variety of nutrients you will obtain from the cured humus. The reason is understandable. Some plants take up more of one nutrient than another. The nutrients absorbed are transmitted throughout different parts of the plants; in roots, fruits, stems, leaves. It follows that the utilization of the roots and stems of some plants, the leaves and stalks of others, will add different amounts of different elements. So much the better for your compost and soil improvement program!

MANURES MAKING COMEBACK NATURALLY

Livestock manure has great value, whether it is used to garden or farm organically as fertilizer or simply added to compost. Of interest to all gardeners is the increasing attention being paid to the value of organic techniques. Even those who have in the past shunned organic gardening methods and otherwise misused their soil are now beginning to realize that there is a rhythm and reason to nature.

Whichever compost system you choose, you can always find ways to improve upon it with ingenuity. One thing is certain: Good compost makes good gardens. If that's what you are after, start composting right now and stick with it! Compost truly is the golden key to more productive, fruitful gardening.

MULCH ALSO HAS MAGIC

Mulch is a naturally good idea. Just count the reasons to be grateful for this natural discovery. Mulch preserves moisture, prevents weeds, improves soil condition, avoids erosion, cuts down disease problems, and adds organic matter to the soil. Mulch is an organic gardening bonanza. Next to composting, mulching is the most vital, single natural gardening activity. Most of the needed, necessary mulch materials are readily and easily available, wherever you live. Most are free. The list of materials is endless. Mulch truly is a simple, effective, and practical way to begin a recycling of nature's bounty for better gardening. Once you try it, you'll agree with millions of gardeners that mulching pays in many ways.

MANY MULCH MATERIALS

Consider for a moment grass clippings. leaves, chopped or ground brush and twigs, pine needles. Basically a mulch is any organic material you can find as a protective covering for the soil. That also includes straw, hay, ground corncobs, peat moss, sawdust, shavings, composted refuse you or neighbors make, and even gravel, sand, or stones. It also includes peanut hulls, ground bark, redwood chips, layers of newspapers covered with grass clippings, and whatever will decompose to add nutrients to the soil.

Yes, black plastic that you buy for weed control in garden centers and similar available coverings can be used as a mulch. So can other materials that stop weeds, help soil retain moisture, and save you weeding time, such as aluminum foil, layers of newspapers, and any other artificial ways to thwart weeds. However, consider appearance. Most important, consider your goal: improving the soil and growing ground beneath any mulch.

Focus first on organic materials that break down and recycle into the good earth. Whatever material you use and apply depends, of course, on what you have available and what works into your budget and gardening goals. If you like to see dark, rich-looking soil around shrubs and trees and along garden rows, you can opt for peat moss, ground sphagnum moss, or chipped bark and wood chips. Well-decayed sawdust is darker. Redwood chips and pine bark look nice. There are drawbacks and basic problems with some materials, however. Wood chips and cellulose materials draw nitrogen out of soil, so you must compensate for that by adding higher nitrogen fertilizer to your garden ground.

Dry lawn clippings are fluffy when first spread. You can gather them abundantly from baggers on your power mower. After a few rains, they tend to form a compact, thin layer. They do decompose and you can add more layers as the season progresses as you mow your lawn each week. The decomposition has its benefits, of course, by improving your garden soil.

Mulch, whatever you use to cover the garden ground to suppress weeds and conserve soil moisture, has its advantages. However, we should think of mulch as a long-range goal to improve garden soil for garden performance improvement. By smothering weeds with mulch, we also preserve soil nutrients and the fertilizer we apply so they can feed our desired plants.

Controlling weeds is an annual, tiresome chore. True, you can pull or cultivate or till them away, but some weeds are frustratingly persistent. They also leave seeds in the ground, just waiting for you to till so that the seeds can sprout, grow, and rob your plants of the nutrients and moisture they need. Also, excessive or overly deep rototilling can damage the spreading underground roots of your valued plants. Weeds also may be hosts to harmful insects that will cross over from weeds to attack your vegetables and flowers. Happily, proper mulching stops weed growth effectively. It smothers the seedlings, prevents other seeds from sprouting, and encourages those useful underground allies, earthworms, to work their wonders in the soil around your garden plants.

Hard rains can cause erosion on bare ground. The problem is worse when you cultivate too often with a hoe, weeder, or rototiller. A mulch takes up the shock of pounding rain, letting it seep into the soil below more evenly, and then holds that water by preventing evaporation from the sun. A mulch also discourages a hardpan, that undesirable baked top layer of soil, from developing on the surface in dry periods.

CONSIDER SOIL TEMPERATURES, TOO

Some plants prefer warm, sunny weather. Others favor cooler conditions. The use of mulches, together with proper planting and culture techniques, can help you extend your gardening season by adjusting soil temperatures. You can prove that point to yourself. Place a thermometer an inch or so deep beneath a mulch of grass clippings or leaves. The temperature may register 75 to 85 degrees. Then place the thermometer in the same patch of soil and remove the mulch for 3 or 4 feet around it. You'll notice that the temperature may climb 10 to 15 degrees. Next, place the thermometer on a block of wood on top of the soil. The temperature may climb to 100-plus. That high temperature can be stressful to many plants. In effect, mulch helps you regulate and balance soil temperature to the desires and needs of your plants. You are, in effect, helping to better control the microclimate around them. Remember that light-colored materials such as pebbles, gravel, straw, and light-colored sawdust, will act to reflect light. Darker materials including peat moss, pine bark, compost and humus will absorb light and heat.

Timing is important with mulching. Many natural gardeners prefer permanent mulching. This is basically like composting right in your garden. As mulch materials continue to rot, they give the added advantage of continued humus production. To plant flower seedlings, or bulbs, merely move the mulch to one side and plant what you wish. Then, respread the mulch and it is back in place to smother weeds, preserve soil moisture, and let your valuable plants grow better without weed competition for space, water, or nutrients.

MAKE MULCHING A HABIT

When seedlings are up and growing well, apply mulch around them to thwart weeds. At the end of a garden season, you can elect to leave mulch in place or till it under to further improve soil texture, condition, and fertility. Because mulch materials may not be readily available in large quantities, selective mulch fits into most garden plans better. Either way, mulching pays big growing dividends as it saves you weeding time.

Compost is your best mulch. As you create compost and humus, this decomposed organic matter provides an attractive and more useful mulch because it already contains some nutrients, but also through the composting process that generates high temperatures in compost piles, many weed seeds are killed.

Keep one thought in mind with all types of mulch. It is usually necessary to add a nitrogen source to organic mulch, especially when you use wood chips or similar material high in cellulose fiber. It can be added before mulching, of course. But as mulches decay, they tend to take some nitrogen from the soil for this natural process. It is up to you to keep things in balance. Wood shavings and wood chips especially require extra nitrogen to assist in decomposition and in maintaining the balance of fertility you need for optimum plant growth. For organic gardening, blood meal, rotted manure, fish emulsion or fish meal, cottonseed meal, and similar materials are good organic nitrogen sources. Add one to two cupsful per bushel of mulch.

Naturally, you can use conventional commercial fertilizers that provide various percentages of nitrogen in them. Read the labels on fertilizer containers to be certain you are using the proper fertilizer formulas for the type of crops you plan to grow. Ask your garden center specialists about the many types of fertilizer available today. They are often your best source for useful information on new garden products that are being introduced to help us grow better.

Remember that too much nitrogen leads to vegetative, leafy growth. For best blooms from flowers, you must be sure they get their needed amounts of phosphorus and potash too. Fortunately, the handy new slow-release fertilizers, prilled and pelleted types give season-long feeding to plants. Popular new liquid fertilizers, applied in hose-end sprayer units, are another modern blessing for gardeners everywhere.

LANDSCAPES BENEFIT FROM MULCH

Your permanent landscape shrubs also gain from mulches. Roses, perennial flowers, and flowering shrubs grow better when well mulched. Azaleas, rhododendrons, and similar acid-loving plants benefit from acid-producing mulches such as oak leaves, pine needles, and similar material.

Even though mulch is an important asset, don't get carried away. Apply only a sufficient layer to prevent weed growth in summer. Too much mulch may retard water from penetrating properly to the soil if the mulch compacts. Your best rule of green thumb is to apply only thin layers of dense mulch at a time and add more later as the early mulch decomposes.

Leaf mold, peat moss, sawdust, and wood chips can be applied to a depth of just a few inches. Grass clippings, old mulch hay and straw, and other types of fluffy mulch can be applied 6 to 8 inches thick. Eventually, of course, all mulch materials will shrink down as they are soaked by rain or irrigation water and begin decomposing into humus. When fall arrives, you can elect to leave mulches in place and just add more next year to provide an ongoing field composting system right in your garden. If you can take a drive into the country, check with farmers. Spoiled hay is often sold cheaply as mulch hay. A few bales of hay can cover an area 15 by 15 feet.

Peat moss is probably the most readily available and common mulch. Best bet is to buy the biggest bales or bags for cost efficiency. Although peat moss has no nutrient value, most other mulch materials do. For example, lawn clippings provide about 1 pound of nitrogen and 2 pounds of potash for each 100 pounds of dry clippings. Researchers also have found that leaf mold or shredded leaves can provide a nitrogen content as high as 5 percent. Alfalfa or clover hay are higher in nitrogen than orchard grass or timothy because these legumes have nitrogen-fixing bacteria in their root nodules. That is why farmers prefer alfalfa or clover hay to provide a high protein value for their livestock.

OTHER MULCH POINTS

As much as mulch deserves a place in your gardening plans, there are some commonsense precautions to observe. Don't mulch the soil when it is excessively wet. Molds can start below the surface and the trapped moisture combined with heat can cause hidden mold problems. You may also contribute to damping-off diseases on seedlings if you mulch too early before they have true leaves and a strong roothold. Gravel, stone chips, or even flat stones can be used as a substitute for organic mulch. They provide some weed control, stepping-stones, and hold early season heat in the soil.

As you plan your garden, think mulch. It is a time saver when you consider how much time it takes to weed your gardens, especially if you don't get them early and they begin to spread. The many other values of mulch make it a key contributing ingredient to your gardening success story.

KNOW THE ABCS OF NPK, TOO

All gardeners should adopt the use of compost and mulch as basic good growing practices. Naturally, there is another key consideration. Just as we must eat properly to live well, plants need nutrients for their growth too. That is as true for flowers as it is for vegetables, shrubs, trees, and other plant life.

It pays to understand the big three—NPK—of plant nutrition. They stand for N—nitrogen, P—phosphorus, and K—potassium. Knowing the part these major elements play in plant vitality and performance is useful to help you get the most glorious blooms and give all landscape plants the nourishment they need for peak performance.

Nitrogen (N) is the key element for vegetative growth. This element promotes strong and healthy leaves, stalks, and stems. In fact, it is vital for all green leaf tissue. Nitrogen fosters development of proteins, the cell-growth builders in your plants. Without this essential element you'll see yellowed foliage and stunted growth. Plants become weak and susceptible to disease, especially under adverse wet or dry weather conditions.

Too much nitrogen also can cause problems, just as overeating does for people. An oversupply of nitrogen encourages excess leaf and stem growth at the expense of flower and fruit formation. Some people feel that if X amount is good, they should double or triple that for better results. Not so! Another problem is using lawn fertilizer on flowers. It is high in nitrogen, such as 10-5-5 or 20-10-10 on various formula labels, so it feeds grass well, but the higher nitrogen in lawn fertilizer formulations is wrong for flowers and vegetables too. That's why you see many different formulations of fertilizer today in all brands. They are designed and formulated for the specific plants, whether lawn, shrubs, fruit trees, or flowers.

Phosphorus (P) is vital for strong, prolific flower development, good fruit set, and seed production. Phosphorus also is required for proper development of plant sugars during the growth of plants, especially berry and tree fruits.

A lack of phosphorus is easily noted. Plants will be stunted and have a yellowed look. This may appear remarkably like nitrogen deficiency, but look again. If there is a distinctive purplish color around the edges of leaves, it usually signifies phosphorus deficiency.

Equally important, through unseen, is the retarded root development when phosphorus is lacking. Leaves may fall; plants may fail to flower. That's when things are really bad with your phosphorus supply in the soil. Plants tell you when they are in trouble.

If phosphorus is out of balance with other elements, superphosphate—available from garden centers—often can rescue crops. It contains more phosphorus per bag.

Don't ask why K stands for potassium—the chemists say so! Actually K_2O is the chemical expression for potash, for practical purposes the same as potassium. This basic ingredient promotes strong, healthy roots. Potassium or potash also helps in seed production. More important to you, it quickens maturity of plants and may help in disease resistance.

A deficiency in potassium is marked by yellowish mottling. In severe cases, foliage loss occurs and roots won't develop well. Also, the formation of fruit is poor when potash is low. That's important if you have berry bushes, fruit trees, or flower shrubs that provide fall berries for attractive appearance and as food for birds.

Those are the essential fertilizer ingredients. They are available in varying ratios, as noted on the bag. They will always appear on fertilizer bags, in fact, in the same order: N-P-K. Translated to numbers, 5-10-5 means your bag of fertilizer contains 5 percent nitrogen, 10 percent phosphorus, and 5 percent potash. Translated still more, a 100-pound bag of 5-10-5 will have 5 pounds of nitrogen, 10 pounds of phosphorus, and 5 pounds of potash.

You can get high analysis material, such as 20-20-20 and other concentrated formulations which contain greater quantities of nutrients per bag. The other pounds in a bag are called carriers, which are inert materials that help you achieve even application of all elements when spreading by mechanical means or by hand. Too much of any element can burn foliage and injure plants.

MODERATION IS GOOD PRACTICE

We Americans tend to overdo things. We eat a bit too much for our own good at times, especially our favorite foods. We also tend to overdo our pesticide applications. Too often when we see a few insects, it has been a bad habit to blast them away with spray. Some novice gardeners seem to believe that if a little fertilizer works well, much more will give even better results. That's just not true. It is essential to read and heed directions on labels, of pesticides and fertilizer materials too. That gives us pause to think and act in moderation, a wise practice in our gardening activities.

Moderation works well in flower gardens too. We should focus on applying fertilizer in smaller amounts and on a regular basis to feed the plants what they need, as they grow, mature, bloom, and set their roots and systems for the next year's growth too. Flowers and other plants also need small amounts of other nutrients, but most soils have these elements in them. Adding compost and organic mulch that decays slowly also replaces small amounts of nutrients year by year.

Commercial fertilizers often also typically contain trace elements in their formulations. Manufacturers know the needs of area soils better than most of us. Commercial fertilizer manufacturers often combine minute amounts of necessary microelements in mixes for areas they know are deficient. You'll find other details about the fertilizer you use on the bag, box, jar, or container. Ingredients must be identified, as required by state laws. You can also find details about application rates and other useful plant nutrient details on many packages, or with folders available when you buy your fertilizer.

As gardening continues to boom, new and improved products are becoming more widely available. There is more information available on containers. Newer, slow-release and high-analysis fertilizers in various formulations are readily available nationwide. Schultz and Miracle-Gro are two popular national brands. Many other manufacturers also provide a range of fertilizer options for different crops: vegetables, fruits, trees, flowers, shrubs, house plants. Manufacturers also are putting more detailed product information plus tips and ideas on the packages and in free flyers at stores. That's good. It makes it easier to select the right nutrient combination plant food to meet your needs.

SPRAY-ON FERTILIZERS

In recent years, much ado has been made about foliar feeding, which is feeding through plant surfaces, and it has proved very successful for those who take the time to do it and follow through. Many different formulations are available. Liquid fertilizer does have a place. Farmers have long piped liquid fertilizer to crops in irrigation water. However, some of the TV commercials tend to make viewers think that all they need to do is point the hose, spray on the plant food, and they'll have super gardening success. Not quite true!

Don't always believe what you see on TV. The commercials are intended to sell you products, so they are dramatic, colorful, and deliberately compelling. Take your time to think about your plants. Learn what they need. Ask local garden centers, nurseries, and other responsible, veteran local garden product suppliers for advice. They are better tuned in to your local weather, soil, growing conditions, and the variety of garden products available today.

Part of the TV commercials are true. Periodic fertilizing does lead to stronger, healthier, bigger, and better plants and blooms. You'll discover that to your pleasant surprise when you begin a regimen of regular plant nutrition. Remember, however, that 10-20-10, or any formula, represents the amounts of nitrogen, phosphorus, and potassium in that particular fertilizer, no matter who makes it or what brand you buy. Some have better formulations for hose-end sprayer application of liquid fertilizer and may have some minor elements that have value too. Basically, the level of nutrition is measured by what is in the formula, which, by state laws, must be clearly stated on the containers. State inspectors spot-check regularly to protect consumer interests.

Another excellent gardening information source is your County Extension Specialists who are backed up by State Horticultural Extension Specialists. These are talented, trained, and dedicated people who are state-federal employees. They are connected with the Land Grant Colleges in their respective states. Their services are paid for by your tax dollars. Tap into their knowledge as you plan, plant, run into problems, or whenever you need help or advice. They are a wonderful group of willing advisors at your call. They also often have a vast store of local knowledge to share.

To be even more certain that you are feeding your Biblical flowers and other plants, as well as your overall landscaping as you should for your soil conditions and their needs, periodic soil tests are a good idea. Think of your soil as a bank. Your plants can only withdraw what the soil has in it for their needs. After that, you must make regular deposits of nutrients for your plant's account. As they grow, prosper, bloom, and set seeds or the leaves rebuild flowers within the bulbs for next year's blooms, plants naturally use up nutrients. It is up to you to maintain a reasonable balance of nutrients for your flowers.

Happily, the many modern fertilizer formulations for flowers, from bulbous flowers to other types, are readily available in convenient packages, labeled for various flowers. Read the package labels, consult the knowledgeable authorities at the garden departments of chain stores and local garden centers, and use what you need for the types of flowers and plants you grow.

UNDERSTANDING SOIL pH

One more key point should be understood about nourishing plants. That is soil pH.

You don't need to be a Ph.D. to understand it. Basically pH is a method for measuring relative acidity and alkalinity in soil. Scientists have devised a simple chart, a scale ranging from 1 to 14 called the "pH scale." Acidic soil registers below the 7 mark, alkaline soil above that point. Most soils in Eastern states tend to be more acid than Western soils.

Acid soil is often called sour soil. Actually, many plants prefer soil slightly on the acid side. That is because slightly acid soil helps break down both basic and minor nutrient elements in soil more easily so that they can be absorbed into plants more readily. To be picked up by plant roots, nutrients in the soil must be in solution.

Regular soil testing, with easy-to-use kits you can obtain from your local garden supplier, let you determine what your soil pH is. Be sure to check different areas. You may be surprised at the pH, and also the NPK—the nitrogen, phosphorus, and potassium—differences in different parts of your home grounds, especially in moister areas. County Extension Agents, your reliable garden advisors in each county in America, are also helpful in providing tests and advice for improving your soil's condition and nutrition.

LEARN ABOUT LIME

Lime is not actually considered a fertilizer, but it is important for adequate plant performance. It sweetens overly acid soil. Bags of lime are inexpensive, and widely used to improve soils for growing lawns. A few bags may be all you need to improve overly acid soil for your Biblical flower garden areas, if necessary.

Flowers need phosphorus, but phosphorus is somewhat tricky. When soil pH is high, on the alkaline side, calcium deposits can lock up phosphorus into a calcium-phosphate compound which makes it useless and unavailable to the plant. When you lower the pH, phosphorus is released. On the other hand, in acid soils, low on the pH scale, iron locks up the phosphorus. So, a soil pH balance of moderation must be maintained.

If some of this sounds difficult, you have an army of garden advisors and helpers available. Local garden centers and the garden areas of major chains often now have garden specialists who can give advice, answer questions, and guide your growing programs. Also check with longtime nurseries and private garden centers that have deep roots in the community. They are more likely to have extensive local knowledge of soil, growing conditions, and also sources for a much wider variety of plants that chain store clerks can't provide.

MORE GARDENING KNOW-HOW

In addition to the information in this and other chapters, you'll find many sources for free catalogs, good growing advice, websites, and links to a wealth of know-how listed in Chapters 11 and 12. County Extension Agent offices also have free gardening brochures on a variety of topics. They are listed in your phone book and their advice is free because they are state and federal employees. So is each State Extension Horticulturist, usually located at the State College of Agriculture. The U. S. Department of Agriculture also has many free and low-cost brochures and booklets.

You'll find specific details about planting and care of various Biblical plants in this and Chapter 2, as well as Chapter 3 about growing the wilder flowers rooted and native that give their glory to the Holy Land today as they have for centuries. As gardening has become America's Number 1 family hobby, it seems logical to tap the vast store of knowledge available for all aspects of gardening. In addition to Biblical flowers, you could possibly have trees, shrubs, herbs, and vegetables growing in your home grounds, or planned for future garden projects. Therefore, I've also tried to provide a much wider range of resources for your use in this book. Use them as you wish as you expand your gardening horizons. `

I've been happily gardening for more than 50 years, beginning in my 4-H Club days when I was just 10 years old. Through the decades I've learned much from experts. As a nationally syndicated garden columnist for 25+ years, the author of 50 published books, and a veteran member of the Garden Writers Association of America, I've also been in regular contact with many top horticulturists, specialists, advisors, and experts on all gardening topics. Today, much of that wisdom is also available on websites of major mail-order seed, plant, and nursery firms. You'll find my favorites in this book.

In addition, I'm gladly sharing my list of favorite mail-order firms who offer free and usually colorful, illustrated catalogs.

More of them also include good growing advice these days. Even better, you'll find toll-free telephone numbers to call to get answers to your questions from real human being gardeners with know-how to share.

Gardening today is easier, more fun, and rewarding because so many people share an interest in making the good earth bloom and bear abundantly. Tap into the resources and sources available and you, too, can enjoy God's wonders in the deserved beauty and magnificence of flowers on your home grounds as well as in container gardens in cities.

Chapter Five

❧

Beautify With Biblical Flowerscapes

Beautiful Biblical flowers in beds, borders, and around your home landscape give you colorful delight and add lasting value to your home as well. That's in addition to the uplifting spiritual significance of these glorious plants. In this chapter, you'll discover ways to glorify God and beautify your home landscape with Biblical flowers. Think of that as Biblical Flowerscaping, a colorful new phrase for a meaningful new growing experience. You'll learn some helpful how-to secrets in this chapter.

Most of us expect to enjoy the gardens we grow for years to come, providing we remain in the same home. However, America is a mobile country. Millions of people move each year, because of business transfer, to be closer to other family members, for retirement, and other reasons. It's comforting to know that our gardening efforts will be rewarded in more ways than we may realize. Attractive landscaping actually can increase the market value of a home up to 10 percent, according to recent real estate surveys. For a reasonable cost, a homeowner like yourself can create strong visual appeal that beckons home buyers and sets a warm mood.

Renovating kitchens and baths is important for home resale value, but just $1,000 spent on upgrading your landscaping pays handsome dividends. It grows bigger and better every year, naturally. Blooming flowers in the garden and cut flowers scenting the house when guests visit are important points in presenting a home's best face. They are also a delight we can enjoy ourselves.

GOOD LANDSCAPING BOOSTS HOME VALUE

Research has demonstrated that homes in neighborhoods with overall excellent landscaping sell for about 7 percent more than similar homes in other neighborhoods. All things being equal, improving the quality of a home's landscaping can boost resale prices 8 to 10 percent. Homeowners who upgrade their landscape quality from good to excellent are returned 4 to 5 percent in increased resale prices in neighborhoods where other homes have excellent landscaping. It's nice to know that your gardens reward you in economic ways too.

Planting perennial flowers—which include most Biblical flowers—plus trees and shrubs should be first focus in any landscaping and gardening plan. As you dream about your future gardening projects, it is worthwhile to focus on that key word, landscaping. Actually, I prefer the word "plantscaping," especially as I plan and plot my own Biblical garden plantings and help others with theirs. Biblical flowerscaping is even more appropriate.

Let's do some ABCs of attractive landscaping. Consider how your house looks from a distance, especially as you approach your front door. For improving overall appearance, it pays to plant attractive shrubs along the front. Flowering shrubs offer spring and summer beauty. Use multipurpose trees and shrubs that have blooms in spring and colorful fruits or berries in the fall. Ask your local nurseryman which do best in your area. Use them to frame your house and showcase your gardens. Local landscapers also can suggest attractive plants to enhance the overview of your property. Get to know local garden experts and sources, especially at nurseries and private garden centers; they'll have the experience to know which plants grow best in their area. Equally important, getting to know good local gardeners leads to new friendships in your shared gardening hobby.

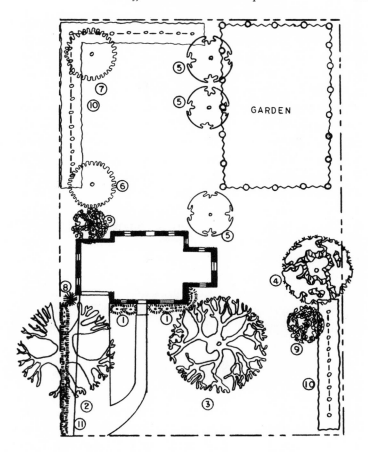

As you plan your Biblical garden, plot out where you have or want trees, hedges, vegetables, herbs, and fruits. Here is one simple design as a basic property plan idea.

1. Biblical flowers providing a colorful welcome to your home.
2. A tall oak shading one corner of the home.
3. A present tree marked for its mature growth area.
4. Another existing tree at the property side line.
5. Apricots, representing trees of the Scriptures.
6. Hawthorn, another tree of the Holy Land.
7. Another property border hawthorn tree.
8. Juniper at the corner of the house, representing another Biblical tree.
9. Mulberry, another representative Biblical tree.
10. Cedar or arborvitae hedges, representing other Holy Land trees.
11. Flower border of various Biblical flowers along driveway.

Look at your entryway again. Appealing shrubs and flowers say welcome. Biblical flowers, crocus, daffodils, hyacinths, tulips, lilies, give you color and beauty from earliest spring through summer. A bed started now, another added later, some naturalized narcissi under trees next year, all grow on for years.

Most Biblical plants grow from bulbs and are perennials. Plant them well and help them get a strong roothold and they'll grace your property, add their perfume to the air, and provide many years of pleasure. Beauty that grows again each year is another one of God's gifts to all of us. Annuals such as marigolds, petunias, and zinnias are nice for fill-in, but perennials give you the opportunity to build a long-range landscape look that has deep, lasting roots. Many mail-order catalogs today give blooming dates so you can have blooms from different flowers all season long. In Chapter 11 I've included a list of my favorite mail-order firms that offer free catalogs and include Biblical flowers in them. These catalogs provide blooming dates so you can select flowers to give you blooms from the first warm spring weather right up to fall frost.

FOCUS ON TREE VALUES

Colorful landscapes are a welcome delight. Multipurpose trees and shrubs yield double rewards: blooms in the spring and fall berries or attractive foliage color. Pick the brightest colors of the fall foliage trees you prefer at local nurseries each fall. The brilliance of an individual tree's foliage you see in a nursery is a genetic trait and will be retained throughout that tree's life. Select the brightest, mark them, and plant them come spring. Here's a color checklist.

For red fall colors, plant sugar maples, which have red, orange, and yellow foliage. Scarlet oak is scarlet. Sassafras turns orange to scarlet. Red maples will be crimson. Sweet gum is scarlet to burgundy.

For yellow colors try ginkgo, yellowwood, birch, beech, poplar, aspen, and Norway maples. For browns and oranges plant hickory, black oak, hornbeam, white oak, and horse chestnut trees.

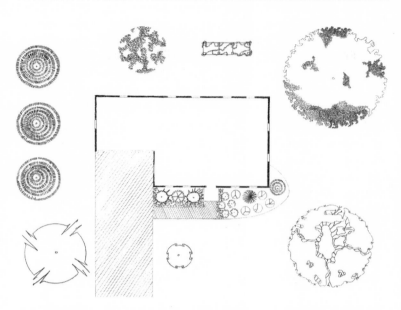

This basic landscape plan demonstrates how easily you can add plants of the Scriptures to your home grounds. The larger trees represent oak, willow, poplar, and cedar as a privacy screen. Pine and apricot can also be planted in the front area with a sweeping willow. Along the walk, spring-flowering bulbs will add their beauty. You may also enjoy grapes on an arbor in your backyard as another favorite Biblical plant.

Odd as it may seem, trees and shrubs can save energy money too. Research has demonstrated that cities can be 5 to 10 degrees warmer than suburbs. This "urban heat island" effect is caused by heat-absorbing surfaces: roads, buildings, driveways, homes. As much as 8 percent of electricity demand for air conditioning is used to compensate for this heat island effect. Considering our nation's ongoing discussion of an electricity crisis, saving energy is more important than most people realize. Trees are nature's air conditioners. By providing shade during summer, they can reduce temperatures of roofs and walls by 8 to 15 degrees.

They also transpire tremendous amounts of water daily and water is the heart of many cooling systems. You can also plant trees to funnel breezes to certain parts of your home grounds.

In other research, shade trees helped reduce air-conditioning costs by almost half compared to unshaded homes in sunny climates. Another study revealed that shaded outside walls were 5 to 10 degrees cooler than unshaded walls. With electric costs rising and supply problems troublesome, trees can be a growing value wherever we live. Pines and cedars are two traditional coniferous trees that keep their leaves year round and have Scriptural roots. In Isaiah 60:13, it is written, "The glory of Lebanon shall come unto thee, the fir tree, the pine tree," and in Psalms 29:5 it is written, "The voice of the Lord breaketh the cedars; yea, the Lord breaketh the cedars of Lebanon." Further, in I Kings 5:6 and 10 it is written, "Now therefore command thou that they hew me cedar trees out of Lebanon . . ." and "So Hiram gave Solomon cedar trees." Coniferous trees have the advantage of reducing winter winds.

DECIDUOUS TREES HAVE VALUE, TOO

Deciduous trees that drop their leaves each fall let the sun warm your home during the winter. That makes other trees with roots in the Scriptures—oak, poplar, willow—even more worthwhile additions to your landscape plan. For example, in Joshua 24:26 it is written, "And Joshua wrote these words in the book of the law of God, and took a great stone, and set it up there under an oak . . ." The willow is mentioned in Hosea 4:13, "They sacrifice upon the tops of mountains, and burn incense upon the hills, under oaks and poplars and elms, because the shadow thereof is good . . ." and the willow in Job 40:22, "The shady trees cover him with their shadow; the willows of the brook compass him about."

Experiments in the Midwest revealed that shelter belts, rows of trees, are actually more valuable than people realized in fuel conservation. In one test, a shelter belt on three sides of a house accounted for a 70 percent reduction in wind and 40 percent

Statues of Mary
and David grace the
Biblical Garden at
Magnolia Plantation
in Charleston,
South Carolina
(Ted Hubert)

The Prince of Peace Bible Garden in Augusta, Maine *(Allan Swenson)*

The dramatic beauty of spring *(Netherlands Flower Bulb Information Center)*

Anemones
(Allan Swenson)

Crocus
*(Netherlands Flower Bulb
Information Center)*

Cyclamen
(K. Van Bourgondien)

Hyacinth
(K. Van Bourgondien)

Daffodils

(K. Van Bourgondien)

Iris,
yellow and purple
(Allan Swenson)
(K. Van Bourgondien)

Lilies, white and orange
(Netherlands Flower Bulb Information Center)
(Allan Swenson)

Poppies
(Neot Kedumim)

Star of Bethlehem
(K. Van Bourgondien)

Wild roses
(Allan Swenson)

Tulips *(Netherlands Flower Bulb Information Center)*

Crown daisies *(Allan Swenson)*

Water lilies
(K. Van Bourgondien)

Hibiscus
(Allan Swenson)

Delphiniums
(K. Van Bourgondien)

Flax
(Allan Swenson)

Sage
(Allan Swenson)

Hollyhock
(Allan Swenson)

For kids
(Netherlands Flower Bulb Information Center)

In your home
(Allan Swenson)

The Biblical Garden at Magnolia Plantation,
Charleston, South Carolina
(Ted Hubert)

savings in fuel. That's welcome news in our increasingly energy-conscious era. Most winds come from northern or northwesterly directions. For years, Midwest farms had windbreaks of trees on north and westerly sides. We can benefit today from the wisdom of farmers of yesterday.

Trees also have another sound benefit. They break up street and traffic sounds to reduce noise pollution. For those who detest the louder, noisier distractions of today's mechanized world, tree planting gained another plus. Highway departments have reported that dense rows of trees and shrubs reduce noise pollution by many decibels, as stated in one recent report that trees along a busy highway can reduce noise penetrating to surrounding areas by as much as 60 percent. That saves your ears and helps ensure peace and tranquility in your gardens. Trees also trap soot and dirt as they filter the air we breathe. Rains wash these residues off leaves back to the ground.

TAKE A LANDSCAPE OVERVIEW

Although this book primarily focuses on flower gardens, it pays to take an overview of total landscaping to provide the best settings for flower gardens and also a more enjoyable outdoor living environment. Trees are a major landscape factor. They were also of major importance in the Holy Land, according to the Scriptures. You can find many mentions of trees in the Bible, including acacia, almond, cedar, laurel, mulberry, oak, pine, poplar, tamarisk, walnut, and willow.

Proper landscape design also should include a focus on easier maintenance. If you avoid a jumble of trees and shrubs, care is easier. You also can achieve greater privacy in this hectic world with more foliage around you. You should also think tastefully when you redo landscapes. Plan to grow fruitfully with bushes or rows of blackberries, blueberries, raspberries, and strawberry beds. They also provide fruit for your table. Nothing beats berries picked right at their peak in your own backyard.

The Scriptures mention brambles, and both Biblical as well as botanical scholars believe that this refers to berry bushes. I agree.

As you expand your growing horizons, plant to grow fruitfully too.

THE FIVE BASIC STEPS TO LANDSCAPING

As you focus on improving landscaping, a review of the five basic steps will be helpful.

1. Develop a list of existing and desired outdoor features.
2. Draw a base plan showing all the existing structures.
3. Outline and sketch in the major landscape features in general.
4. Draw in the desired new landscape features where you want them to be.
5. Revise and finally plot your finished landscape plan on paper.

With a day or so advance planning and plotting on paper, you'll have a useful guide as you plant trees, shrubs, flowers, beds, and borders to give you that happier, more attractive outdoor environment you deserve. An even better way to do this if you are good at a computer is to purchase one of the handy landscaping software packages. Staples, Office Max, and other computer stores have them. With these you can easily copy in your present landscape then add new trees, shrubs, beds, and borders. You can also ask the computer to show you mature sizes of plants so you can see and evaluate how your plan will look when plants reach maturity. Landscape software is another modern blessing for all gardeners.

CONSIDER OUTDOOR LIVING NEEDS

Looking ahead, consider what you want in your outdoor living room. Do you want a patio for entertaining, a children's play area, a sports area, cutting gardens, vegetable and herb gardens, and fruitful plots too? Do you want a quiet meditation or prayer area? Look through garden catalogs and read about the various plants you may like to have. With your base plan in hand, walk your home grounds at different times of the day. Check to be sure that the sun reaches areas where you would like to have flowers or even vegetables and fruits and will shine there long enough each day to satisfy your plants' needs. Remember that some plants need hours of sunlight. Others can prosper with limited shade. Learn the needs of your plants so you will pick the appropriate ones for the areas in which they'll prosper. Garden catalogs today offer much of this useful information about plant needs for sun, water, space, and blooming times, too.

GARDEN FOR THE BIRDS, TOO

Birds are part of God's world and delight us with their songs and antics. Children love to watch birds, especially those that nest nearby and feed their young in birdhouses you erect for them. Many home buyers love to see birds. Select trees that attract songbirds and provide wildlife habitats. Local garden centers and nurseries offer many. Here are some of the best that are recommended by veteran birders and Audubon Society friends. Aromatic sumac, bayberry, firethorn and bittersweet, Virginia creeper, barberry, holly, and American cranberry bush look great and offer abundant food for birds. Living screens of wild roses and mulberries and even blackberries do well as property borders or to hide unattractive areas. All relate to plants that grow in the Holy Land.

For specimen tree plantings of showy trees, consider autumn olive, mountain ash, ornamental crabapple of all types, and dogwoods. Hawthorn is our favorite for hardiness, blooms, and fall berries. It also has Scriptural roots. Ornamental crabapples bear white, pink, or reddish blooms and bright berry crops in fall.

FRAGRANT BIBLICAL PLANTS

One of the sweetest delights of gardens is the delicious fragrance of different types of flowers. Too often this joy is overlooked as

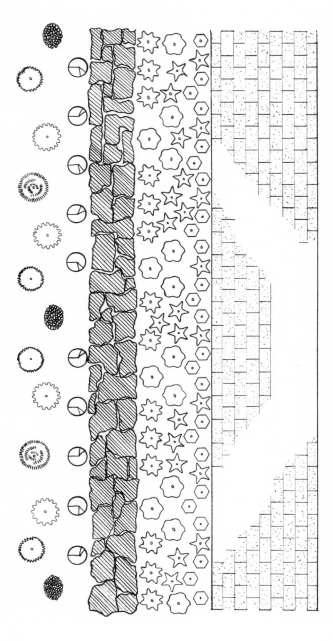

This design can guide you to a combination of flowers and herbs along a path or wall. Lower-growing crocuses and hyacinths can be grouped closest to the path with taller clusters of narcissi and tulips behind them. You can also grow attractive, aromatic, and useful herbs in their own areas among the flowers or grouped together above the wall. By selecting various types of narcissi, tulips, lilies, and iris, you can stretch the blooming season for many weeks, from spring into summer. Although some flowers may not be as closely resemble plants known in the Holy Land, modern hybrid tulips, lilies, and iris share their roots and heritage.

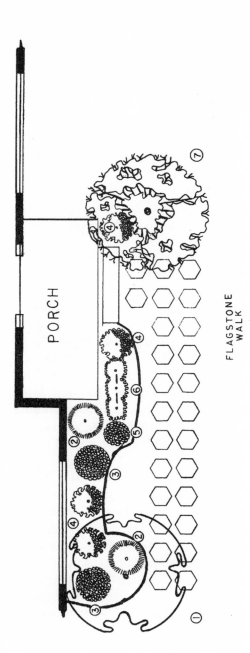

PORCH

FLAGSTONE WALK

You can make your porch and backyard sitting area a delight with Biblical plants, as shown in this basic design. Try a hawthorn, 1 at the corner of your border. Then add clusters of other Biblical plants: 2-iris, 3-lilies, 4-tulips, 5-narcissus, 6-hyacinths, and 7-your choice of representative tree: almond, mulberry, pine oak, whichever pleases you.

we choose flowers and shrubs for color, display, and other purposes. Plant hybridizers have created more beautiful blooms of
modern hybrid flowers but often at the expense of fragrance.
Fortunately, many old-time favorites including Biblical as well
as heirloom flowers of all types still smell good. This year, focus
on fragrance and you'll be pleasantly rewarded every time you
step outdoors to enjoy your garden. Hyacinths are noted for
their fragrance, especially the larger hybrid types. Some lilies
also provide the sweet scents of blossoms. So do some of the
wildflowers of the Holy Land, which I have included in their
own "Wild and Wonderful" Chapter 3.

INHALE THE FRAGRANT FLOWERS

Check mail-order catalogs as you explore the Biblical flower
world and you'll find that more of them now indicate which varieties offer fragrance. For decades, plant breeders competed to
produce bigger and more dramatic roses and other flowers.
They were superbly successful. Unfortunately, they also inadver-
tently often eliminated or reduced the fragrance. Flower-loving
gardeners were not amused. Many wrote to seed companies and
mail-order plant firms to protest the lack of traditional fragrance. Happily, plant hybridizers heard the pleas and down-
right demands of avid gardeners, veterans and newcomers alike.
Fragrance is being bred by crossing heirloom fragrant varieties
back into the flower gene pool. That's the sweet smell of success
in response to good gardener sensitivity.

Some of the flowers you may elect to grow in your Biblical
garden which offer the best fragrance may not be as similar in
appearance as the native species from the Holy Land, or those
from historic days. Through the years, plant hybridizers have
created thousands of different types of daffodils, tulips, and
lilies. Some have marvelous aromas and dramatic, large blooms.
Some varieties have lost their sweet smells. Most don't look
much like the original basic parent plant ancestors that grew in
the Holy Land during the time the Bible was written but their
historic roots are related.

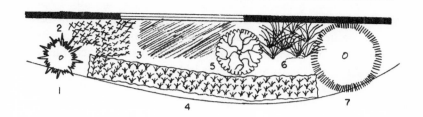

Look out a window and plan the Biblical flower view you want to see. In this design, which you can vary as you like, consider this display: 1-Sargent's juniper; 2-iris; 3-myrtle as a ground cover under the window; 4-bed of crocuses, daffodils, hyacinths, tulips, or a combination interspersed for early to later season blooming; 5-lilies, 6-day lilies, and 7-cedar or juniper.

GROW GOD'S MANY FLOWERS

There are many other plants that you can add to your landscape to impart wonderful, enticing outdoor aromas during their blooming seasons. Although we are focusing on specific Biblical flowers, there is no reason not to enjoy a mix-and-match selection of God's other flowers as we seek to beautify our individual corners of the world. Some plants have perfumes that are subtle; others are fruity or spicy or flood the night with sweet aromas. In celebration of a world that smells better, especially after a day's work in the city, here are some suggestions.

EXPAND YOUR FRAGRANT HORIZONS

You have a wide choice of many fragrant plants to provide the sweetest smells you wish from spring to fall. From lily of the valley to syringa, from clematis to various lilies, from oldtime roses to viburnum. When planting a garden for fragrance, it is wise to concentrate the most aromatic plants near areas where you spend time outdoors; especially in early morning or evening when most fragrant plants are at their peak. Paths, walkways, outdoor dining areas, or quiet sitting nooks are ideal spots. Fragrant plants also deliver their pleasures from window boxes or containers on patios and decks. Many garden centers offer a variety of plants with a bloom period from French lilacs in spring to sweet autumn clematis in fall.

Best bets for fragrant gardens from seeds, according to my good horticulturist friend Lynn Harris, include: carnations, various pinks, mignonette, lavender, fragrant nicotiana, almost all stocks, sweet pea, violas plus ornamental basil, and rosemary. You can also plant biennials that bloom their second year, such as dame's rocket, sweet William, and English wallflower. Two of my favorite catalogs for beautiful, dramatic, and also a selection of fragrant plants are Park Seed, 1 Parkton Avenue, Greenwood, SC 29647, Tel: 864-223-7333, and Wayside Gardens, Hodges, SC 29695, Tel: 800-845-1124. You'll find many other mail-order firms among sources in Chapters 11 and 12 for convenience.

BEGIN YOUR PLANTSCAPE PLAN

As you focus on your landscape horizons, here are key points to remember based on the five basic steps to landscaping I men-

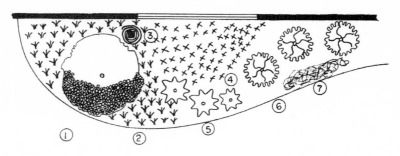

Consider this appealing sight when you look out a window: 1-mulberry, 2-crocus followed by hyacinths, 3-daffodils followed by tulips, 4-tall lilies, 5-day lilies, 6-various iris, and 7-star-of-Bethlehem.

tioned at the beginning of this chapter. I strongly recommend a base sketch plan, locating property lines, buildings and dimensions, major existing trees and structures. Please measure property lines and boundaries. This avoids planting valued trees or plants on a neighbor's land, which I once did in error. Fortunately, we agreed to share my error from the berry patch that was half on my neighbor's sunny land. A replanted berry hedge followed the property line so we would know where our property boundary was and could each pick from our own side of the berry bushes, a peaceful solution to a thorny problem, no pun intended.

Identify All Utilities—Above and Below Ground

Also check with local utilities, especially in cities and new suburban areas. Be sure to locate and mark utilities, both above and below ground. There is nothing worse than locating a new tree where it may look fine, but when you dig you hit a power, water, sewer, or gas line. That has proved to be a dangerous and sad experience at times. If you are having major digging done, make sure your contractor checks for buried electrical and gas lines, for his safety and yours. In some states it is the law.

Know how tall your trees will grow, so they don't send branches into overhead electric lines, including cables to your home. Also know mature breadth or circumference of your trees, so you don't plant too close to buildings and need to cut off branches that rub, scrape, or otherwise cause problems to the house.

Sketch in the approximate locations of major landscape areas. If you plan a patio someday, outline it. Take a careful look at your public area, the way your home looks to friends, neighbors, guests, and passersby. Consider the play or entertaining area behind your home. Does it need screening for privacy or hedges to block out views you don't like? Living hedges, shrubs, and fences work well to screen out unpleasant sights. Thorny bramble berry bushes also make a useful property border to discourage wandering cats and dogs.

Take an Objective Look at Your Home

First step's first as you set your landscape improvement priorities. Drive up to your house and look at it carefully. What does it say to you? Does it say people who care about the neighborhood live there? Does it reflect your good taste in gardening? Does it invite praise? Does the look of your home give you a welcoming feeling? As you look at your home, remember that it is your face to the public, the face that everyone sees first, before they knock on your door. Is it as appealing, friendly, and attractive as you would like it to be? Does it present you the way you would like to be seen and known?

Landscaping should blend your home into the surrounding area so it looks natural and comfortable. To accomplish this, soften the strong vertical lines of the house with trees and shrubs. Trees will frame a house. They can add depth to it. Low homes look best with small- to medium-size trees that won't outgrow and dwarf the house. Bedded shrubs and border plantings create a transition from open lawn areas to the house itself. Attractive plants should lead a visitor's eye to the entrance. You can accomplish this by positioning larger plants at the corners of the house, tapering off to smaller plants toward the door. Taller trees can be used to break the roof line or, in the private backyard area, provide a natural, softer backdrop to your house when viewed from the street.

Next priority, considering how much time you spend outdoors, is your entertaining and living area, perhaps combined with a children's play area where you can watch them while they, too, thrive amid the gardens you create. Hedges and shrubs, as well as fences, provide privacy if that is what you wish. From fast-growing privet hedges to more ornamental flower shrub plantings, you can create areas of privacy, solitude, and prayerful meditation spots. You can add special Biblical plant specimens from mail-order firms, but as you design and plant outdoor living rooms, local flora is better adapted to success.

Consider the Look of Your Total Landscape

As you develop your own expanded landscape horizon this year, some tips from top landscape architects and designers can prove helpful. The advice that follows has been generously given by garden designers and landscape horticulturalists alike. They too share our theme: Growing together, we can create a better world.

There are many different sizes, shapes, and designs of homes that call for different approaches in landscaping. This book isn't intended to be a landscaping guidebook, but key points offered by veteran landscape specialists and designers can help. Here are some of the essential key points to consider, depending upon which type of home you have.

A long, low house calls for careful long-range planning. In a modern contemporary home, taller trees in the backyard and rear of the property provide a natural display of foliage to break up the long, low roof line. Hemlocks, which grow tall and somewhat pyramidal, can be planted to break up and soften lines. They can be pruned periodically to retain a smaller size. Junipers or arborvitae can be used as part of the front shrub plantings. You might also chose alternatives as yews, spreading or upright for the front plantings. Some flowering trees such as almond, hawthorn, crabapple on the lawn can enhance the total living landscape picture with blooms in spring and berries in fall.

Landscaping a sloping entrance should be no problem. A firethorn will thrive at the side of the drive and can be trained to reach over the garage door. Holly, yew, or mugho pine can be used at the doorway with a grouping of shrubs including pyramidal yews at the corner of the home. Taller evergreens from upright arborvitae to cedars might be a choice at the lower driveway level or you can plant medium-size deciduous trees in-

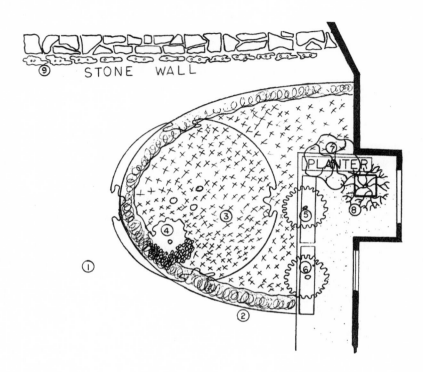

STONE WALL

PLANTER

Try an extension garden beyond your porch to enjoy while reading, relaxing, or entertaining friends. Consider using this design to plant: 1-almond, mulberry, or walnut; 2-border of spring flower bulbs from early crocuses to later daffs and tulips; 3-naturalized narcissi beneath the tree's branches; 4-a laurel or rhododendron bush; 5- and 6-window boxes of whatever Biblical flowers you prefer; 7-burning bush; 8-potted dwarf apricot or apple tree. Also consider using space in the planters for choice herbs to spice up culinary delights.

You can make the entrance to your home a delight to see by combining a variety of suitable trees and shrubs with Biblical flowers as you wish. By planting a variety of spring-flowering bulbs from crocuses and hyacinths to daffodils and tulips, using modern hybrid varieties, you can extend the blooming system for many delightful weeks.

stead. If the slope from drive to front door is steep, consider a rock garden or several terraces. As an alternative, use ground covers like myrtle that hold soil and reduce mowing chores too.

Smaller homes can look mighty small and barren without landscaping. You can create a well-loved look with a few trees and shrubs that won't strain your budget. Intersperse Biblical flowers in beds and borders, at the sides of the front door, and also in a window box or two. Even the smallest houses we saw in Switzerland and France were brought to life with that simple, effective, easy-to-tend, and ages-old garden device, the window box. They are making a colorful comeback in American cities and suburbs alike. Even businesses are dressing up their store fronts with colorful window boxes of flowers.

Trees at the corner of a home extend the size of a small house while giving it a neatly landscaped look. Lower-growing shrubs, from spreading yews and junipers to cotoneaster and similar shrubs beneath windows, draw the eye to the entrance. Beware

A specimen tree, some low-growing shrubs, and flowers in season add an appealing accent to your home as they visually expand its size.

of planting tall growing shrubs that will eventually, and often too quickly, grow up to hide your home. In all landscaping plans, remember that some shrubs require regular pruning. They thrive with it and need it yearly. Other shrubs such as dwarf Japanese yew, rock or spreading cotoneaster, blue rug juniper, are naturally dwarf in growth habit and save pruning time and work. Look at your neighbors' landscapes, ask about their plants, read landscaping books, and dig in. As you do, here's another key tip. Don't overcrowd a landscape plan. Just a few well-chosen trees and shrubs will greatly improve a home's appearance. It is far better to underplant than create an overgrown jungle that hides the beauty of your home itself. Remember, too, that trees and shrubs grow all around, not just upward. Plant them far enough from your foundation so painting and building maintenance can be accomplished properly and easily.

Some final points to help you think about the way you want your landscaping to reflect your personality, your garden viewpoint, and your Biblical plant perspective. A corner planting from a house, with a taller framing tree, helps expand the length of your house from a visual viewpoint. Lower-growing shrubs and flowers can be extended from the house to the tree on the lawn to tie the various plantings together even more effectively.

By varying types of trees with their own unique form, shape, and texture for added appeal, you can give a home an extra-special look. Evergreen and deciduous trees combined with shrubs that have varying growing habits and shapes, different leaf colors, textures and patterns, can often create an interesting and pleasing combination for a long, low ranch home. A taller home usually needs a more massive corner planting and perhaps a large framing tree to set the home off well.

Shape Up Attractively

Shape also has a bearing on landscape design. Some trees are round, others oval, pyramidal, or weeping. Naturally, you will select those that appeal to you, but keep this one point in mind.

Avoid a "plant collection" look with all individual types. It is better to use a few of each to achieve pleasing balance with variations in form, texture, shape, and growth habit. If you are specifically creating a Biblical garden, try to strike a balance where possible, although specimen trees and other plants have their place for their special significance.

The trees traced to Scriptural passages are: almond, cedar, laurel, mulberry, oak, pine, poplar, tamarisk, walnut, and willow. The fruits are apricots (apples), figs, olives, palms, and pomegranates. Unless you live in states with semitropical or tropical climates, most of these latter trees are unsuited for much of America. Naturally, those that were mentioned in Scriptures and still grow in the Holy Land may look different from their relatives that grow and prosper in the United States. For example, there are many different types of oaks and pines. All trace their roots to the same family but look somewhat different. That's natural. Another factor restricts your choice. Cedars of Lebanon and other specific trees of the Holy Land are either not available or so rare as to be prohibitive in cost, or not available for import into the United States.

Know Horticultural Zones

America is a giant country with many different horticultural zones, as the Zone Map on page 146 illustrates. It is far larger than the Holy Land and most of the United States has a four-season climate. Therefore, you are well advised to consult with local nurseries to determine which species will do best and prosper in your locality. As we agreed earlier, God has given us the beauty of flowers and trees and all living things. Therefore, we can dig in and grow those most directly related to the plants of the Scriptures or the Holy Land today, or use our best judgment to grow those most closely resembling or related to such plants.

Some plants like it hot; others prefer cooler weather. Some can withstand stark, frigid northern winters. Others won't survive severe frosts. Your favorite plants from living in another part of the country may remain your favorite tree or shrub, but if it won't survive where you live now, keep it treasured in memory. Local nurseries can advise you which survive and thrive in their area. Take advantage of their well-grounded knowledge.

Horticultural Zone Indicator Plants

Happily, the National Arboretum of the U.S. Department of Agriculture, in cooperation with the American Horticultural Society, has developed excellent plant indicator guidelines. These so-called Indicator Plant Examples serve as a handy reference to typical plants that thrive in the various horticultural zones. The Superintendent of Documents in Washington, D.C., offers a colorful horticultural zone map. It includes details of tree and shrub groups that thrive in respective zones. Local plant suppliers can add to the basic list, based on their experience in the particular area where you live.

The map on page 146 details the expected minimum temperatures in most of the important continental areas of the United States. It is a guide, and local conditions may vary. The indicator plants are based on proven facts as they apply to each zone. There may be, of course, islands of variations within the zones. The map zones and indicator plant lists have been most useful in determining minimum temperature survival of plants in specific areas. For example, much as I loved the dogwood trees that graced our several homes in New Jersey, they just don't survive winters in Maine. Nor do the old-fashioned magnolias that we could grow in the Garden State, about the northernmost part of their range it seems.

You can improve soil condition, fertility, water, and cultural methods, but you still are confined by the normal weather where you live on this planet. With those cautions in mind, on page 147 is the Indicator Plant List with minimum temperatures for your reference.

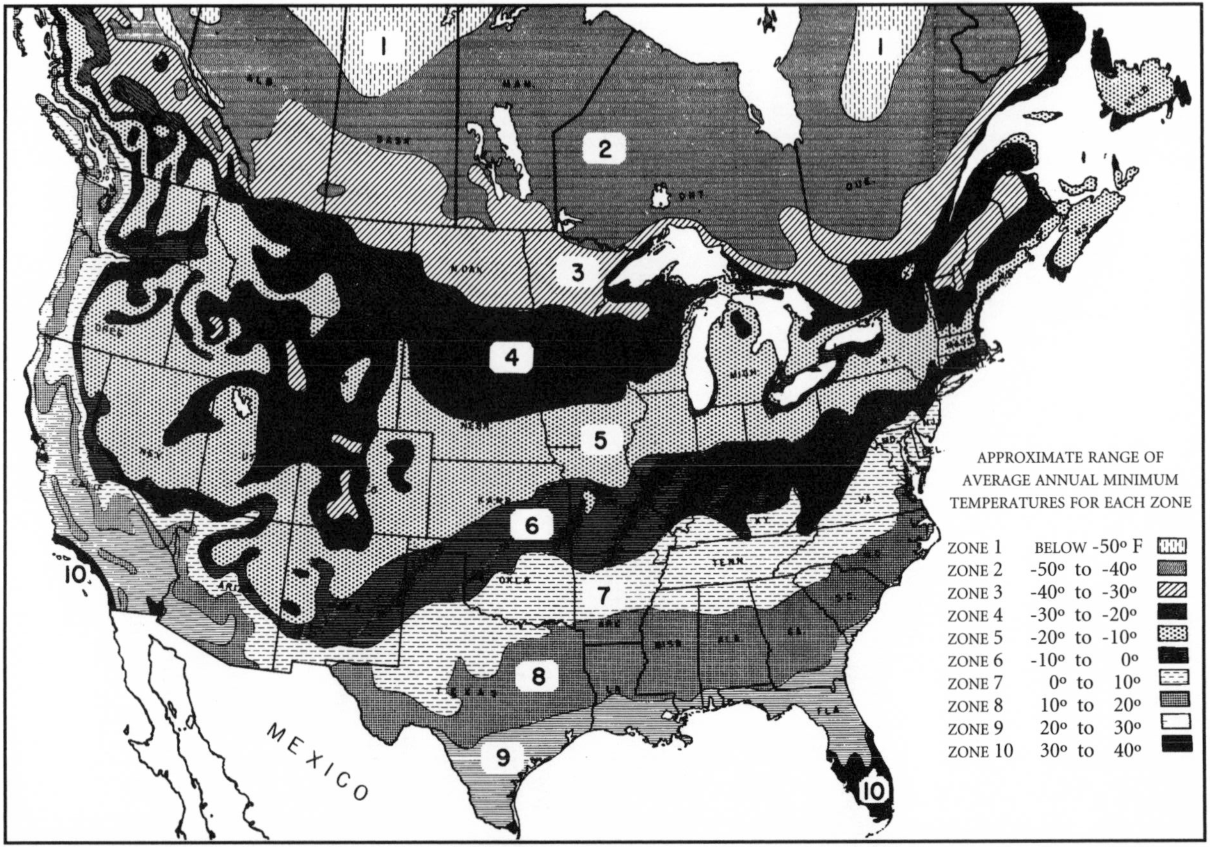
1
1
2
3
4
5
6
7
8
9
10
10
MEXICO
APPROXIMATE RANGE OF
AVERAGE ANNUAL MINIMUM
TEMPERATURES FOR EACH ZONE
ZONE 1 BELOW -50° F
ZONE 2 -50° to -40°
ZONE 3 -40° to -30°
ZONE 4 -30° to -20°
ZONE 5 -20° to -10°
ZONE 6 -10° to 0°
ZONE 7 0° to 10°
ZONE 8 10° to 20°
ZONE 9 20° to 30°
ZONE 10 30° to 40°

Horticultural zones across the United States.

Zone Number	Common Name	Botanical Name
ZONE 1		
(Below -50 degrees F)	Dwarf birch	*Betula glandulosa*
	Crowberry	*Empetrum nigum*
	Quaken aspen	*Populus tremuloides*
	Pennsylvania cinquefoil	*Rotentilla pennsylvanica*
	Lapland rhododendron	*Rhododendron lapponicum*
	Netleaf willow	*Salix reticulata*
ZONE 2		
(-50 F to -40 degrees F)	Paper birch	*Betula papyrifera*
	Bunchberry dogwood	*Cornus canadensis*
	Silverberry	*Elaegnus commutata*
	Eastern Larch	*Larix laricina*
	Bush cinquefoil	*Potentilla fruitcosa*
	American cranberry bush	*Viburnum trilobum*
ZONE 3		
(-40 to -30 degrees F)	Japanese barberry	*Berberis thunbergii*
	Russian Olive	*Elaeagnus augustifolia*
	Common juniper	*Juniperus communis*
	Tartarian honeysuckle	*Lonicera tatarica*
	Siberian crabapple	*Malus baccata*
	American arborvitae	*Thuj occidentalis*
ZONE 4		
(-30 to -20 degrees)	Sugar maple	*Acer saccharum*
	Panicle hydrangea	*Hydrangea paniculata*
	Chinese juniper	*Juniperus chinensis*
	Amur River privet	*Ligustrum amurense*
	Virginia creeper	*Parthenocissus quinquefolia*
	Vanhoutte spirea	*Sopiraea vanhoutei*
ZONE 5		
(-20 to -10 degrees F)	Flowering dogwood	*Cornus florida*
	Slender deutzia	*Deutzia gracilis*
	Common privet	*Ligustrum vulgare*
	Boston ivy	*Parthenocissus tricuspidata*
	Japanese rose	*Rose multiflora*

Zone Number	Common Name	Botanical Name
	Japense yew	*Taxus cuspidata*
ZONE 6		
(-10 to 0 degrees F)	Japenese maple	*Acer Palmatum*
	Common box	*Buxus sempervirens*
	Winter creeper	*Euonymus fortunei*
	English ivy	*Hedera helix*
	American holly	*Ilex opaca*
	California privet	*Ligustrum ovalifolium*
ZONE 7		
(0 to 10 degrees F)	Bigleaf maple	*Acer macrophyllum*
	Kurume azaleas	*Azalea Kurume hybrids*
	Atlas cedar	*Cedar atlantica*
	Small-leaf cotoneaster	*Cotoneaster microphylla*
	English holly	*Ilex aquifolium*
	English yew	*Taxus baccata*
ZONE 8		
(10 to 20 degrees F)	Strawberry tree	*Arbutus unedo*
	Mexican orange	*Choisya tenata*
	New Zealand daisy bush	*Olearia haastii*
	Japanese pittosporum	*Pittosporum tobira*
	Cherry laurel	*Prunus laurocerasus*
	Laurestinus	*Viburnum tinus*
ZONE 9		
(20 to 30 degrees F)	Asparagus fern	*Asparagus plumosus*
	Tasmanian blue gum	*Eucalyptus globulus*
	Bush Cherry	*Eugenia paniculata*
	Fuchsia	*Fuschia hybrids*
	Silk oak	*Grevillea robusta*
	California pepper tree	*Schinus molle*
ZONE 10		
(30 to 40 degrees F)	Bougainvillea	*Bougainvillea spectabilis*
	Golden shower	*Cassia fistula*
	Lemon eucalyptus	*Eucalyptus citriodora*
	Rubber plant	*Ficus elastica*
	Banana	*Musa ensete*
	Royal palm	*Roystonea regia*

Chapter Six

❧

Welcome God's Beauty Indoors in Colorful Container Gardens

Everywhere you look today, container gardens are sprouting, blooming gloriously, and bearing abundantly in pots, tubs, barrels, window boxes, and other decorative containers on porches, balconies, patios, and even rooftops.

As a veteran nationally syndicated garden columnist for 25+ years, I've always watched for and focused on trends that will be of interest to my readers. Container gardening is one of America's newest, fastest-growing gardening horizons, according to recent surveys and reports among the garden-writing fraternity. In Europe, container gardening has been widely popular for decades. Now, this fashionable growing method is being enthusiastically transplanted to cities and suburbs, coast-to-coast here. People are enjoying glorious displays of flowers that lend themselves to small spot and pot cultivation. Biblical flowers do well in container cultivation and will reward you with glorious blooms when you satisfy their specific cultural needs. You'll find out how in this chapter.

We surveyed expert gardeners in Ireland, Great Britain, and Europe, plus nationwide around the United States, about the container gardening trend during our travels in the past few years. Actually, container growing is an old-time European tradition. We marveled at the extent of container flower gardens gracing windows and balconies of homes, apartments, and office buildings throughout Switzerland, Bavaria, Austria, France, and England. Horticulturists estimate that Germany alone has 25,000 miles of container gardens. You can enjoy that blooming beauty with Biblical flowers wherever you live. Ideas, tips, and advice we harvested are included here for your enjoyment.

Comments from savvy horticultural specialists are worth reviewing as they also apply to Biblical flowers in containers today. Far-sighted horticultural authorities had seen the container gardening trend sprouting in America some years ago. Mississippi horticulturist Jim Perry had noted, "There is a definite increase in container gardening in tubs, boxes and windowsills. This is especially true for people in apartment complexes." Don Lacey, a New Jersey horticulturist, reported wide use of conventional redwood tubs, ceramic pots, and barrels. He also reported many improvised containers of chimney and flue tiles, cinder blocks, buckets, and baskets. From Connecticut, Dr. Edwin Carpenter had reported, "I would expect that container gardening will continue to gain popularity, all across America."

WIDE CHOICE OF CONTAINERS

Your choice of container is limited only by your imagination. Wooden barrels, decorative ceramic tubs, pots and planters, and hanging fiberglass and plastic baskets and buckets. Old-fashioned window boxes are making a comeback. You can use anything that holds soil mix and provides proper water drainage. Garden centers and chain stores provide a wide range of shapes, sizes, and colors. Some are functional, others decorative. You can also shop flea markets and yard sales for old baskets, barrels, kettles, buckets, and even fancier containers. Many flowers, Biblical ones and other types, too, will grow well in containers that are 8 to 10 inches in diameter and 8 or more inches deep. Taller plants will need larger pots, tubs, or barrels. Match the container to the mature plant size.

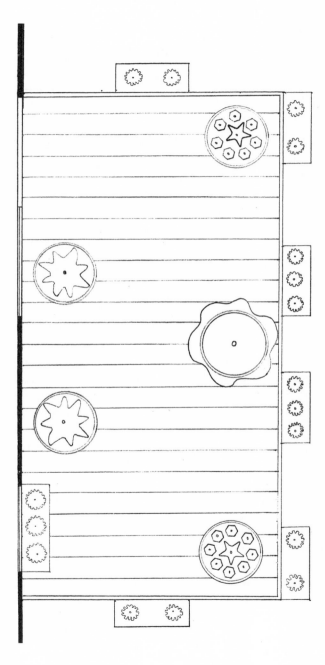

City residents can enjoy their own miniature Biblical garden by growing flowers in window boxes, tubs, barrels, planters, and even smaller pots. Two key points are proper drainage and a porous soilless soil mixture, readily available at garden centers in chain stores today. Then, pick the flowers you prefer, plant them well, and enjoy your blooming beauties from spring through the entire summer by growing a succession of flowers.

WATCH WATER NEEDS CAREFULLY

As you plan your own container gardening, keep plant water needs carefully in mind. Container plants need more water than those in backyard gardens because of their restricted growing habitat. Their roots cannot roam in search of moisture. They also are exposed to drying air on all sides from sun and heat radiated from building walls. Periodic light watering is important to keep plants thriving.

However, proper drainage is especially important. Plants hate wet roots. Be certain your containers provide for drainage and escape of excess water. Place several inches of coarse gravel in the bottom before adding soil mixture. It also pays to use containers with built-on saucers or those that feature newer type wick watering systems. Check with your favorite garden centers to find some of the attractive plant containers with easy-care features.

It is important to check container-grown plants daily to be sure they are getting the water they need, especially smaller ones that can dry out faster. Poke a finger into the soil. If it remains dry, water well. Container gardens need deep watering to ensure that moisture gets to all parts of the soil mix. This encourages deeper rooting for sturdier plants. Avoid evening watering because moisture left on foliage plants, especially in warm water, invites plant diseases. Some modern containers provide moisture on a regular basis with their own bottom trickle or wick watering systems. Ask about the newest containers at your favorite garden center or nursery.

Experts stress, and we fully realize the need for, proper drainage in containers so we emphasize that point again. We always place an inch or two of gravel in a container, especially if it has no or poor provisions for draining excess water out. In pails, buckets or baskets lined with heavy plastic bags to hold your growing medium, allow some holes whenever possible to let that excess water escape. Plants don't like—in fact they can't stand—wet feet. It can stunt their growth, because plant roots need oxygen as well as moisture for proper growth and transportation of nutrients for the best plant growth.

Double potting can solve the problem too. Simply pot your plants in clay pots or tubs. Then place them in a larger, more decorative container. Fill the space between the containers with sphagnum moss or peat moss or sandy gravel. Since the inner clay pots "breathe," they let excess water escape. Keeping the material between the containers moist has beneficial effects, providing moisture during warm days, without clogging soil with excess water.

USE "SOILLESS" SOIL MIXTURES

Another key to container growing success is "soilless" soil. Actually this isn't a new development. Professional growers have used so-called "soilless" soil with exceptional results for years. They call it the ideal plant growing medium. Simply, it is a well-balanced blend of peat moss, vermiculite, which is expanded mica and perlite, a manufactured filler product with plant nutrients added. Most of today's prepared planting mixes are based on the original Cornell University formula. That was a basic peat/vermiculite/perlite planting medium used by commercial plant producers and in greenhouses. Various formulations are available from various planting mix manufacturers today.

One ingredient is long-fibered sphagnum peat moss that is loose and spongy so it absorbs water without becoming soggy. That allows oxygen to reach plant roots. Another ingredient is vermiculite. This is heat-treated mica, expanded 20 times its original volume. This, too, retains moisture. In addition, it attracts plant nutrients while ensuring good water and oxygen flow. Perlite adds even greater fluffiness so roots can grow well and absorb dissolved fertilizer nutrients as well as obtain oxygen.

Today, you'll find different brand names of soil mixtures available in garden centers, supermarkets, and chain stores. They have important advantages over typical backyard earth, which has soil bacteria, fungi, and insects in it. It can be too

Raised beds, tubs, and planters all provide good growing areas for Biblical flowers, even in small backyard spots, patios, and balconies.

sandy or overly wet. The new container soil mixtures are scientifically prepared to give plants their optimum growing conditions. Jiffy Mix is one that works well. Many other brands of prepared container "soil" or growing medium are now generally available. Major chain stores have them available.

GIVE PLANTS BALANCED FEEDING

Plants need a balanced diet, especially in containers where roots can't roam in search of food. Fortunately, many balanced fertilizer mixtures are available for container plant feeding. You have a choice of liquid fertilizer, prilled type or slow-release pellets. Liquid fertilizer allows you to mix the plant food with water, as directed on the package of the fertilizer formula you use.

Prilled or pelleted fertilizers contain plant nutrients in prills or pellets. These dissolve over time so the plants are fed nutrients gradually over a period of weeks or even months, based on the type of fertilizer you purchase.

Fertilizer manufacturers provide useful details on their packages that tell you how much to use for which types of flowers, vegetables, shrubs, and trees. Always select flower fertilizer formulations for your Biblical flower gardens. Flowers need a balance of nitrogen for their vegetative growth of stems and leaves plus phosphorus and potash to form their flowers and bloom as gloriously as they can. Always read and heed the directions for the type you use. Too often Americans have a bad habit of thinking that if a little is good, more will be better. That's not a good idea in any type of gardening, container gardening in particular because of the special restrictions on plants in the containers. It's a fact that too much fertilizer can be as harmful as too little, especially in container habitats.

PICK BEST PLANT VARIETIES

Many types of Biblical flowers can be grown in containers. In fact, container culture allows you to grow many Biblical plants beyond just flowers, including Biblical herbs and even vegetables. I focused on those in my *Herbs of the Bible*, a sequel to this book. For this book, I've concentrated on providing ideas, tips, and advice for growing Biblical flowers. In this chapter, you'll find details and advice for growing them in containers, wherever you live. There's more good news for those without backyard garden space, too.

Considering the fact that millions of Americans want to garden, but live in apartments and condos with no outdoor garden space, the growing container trend this past decade has been most welcome. Fortunately, plant breeders have focused on producing special compact flower and vegetable varieties for containers so everyone can exercise their Green-thumb inclinations, even those in city high-rise buildings. This book is about Biblical flowers, but I wanted to point out in this chapter that container gardening offers far wider opportunities than just with Biblical plants. You can learn much more about container gardening in specialized books about that subject. With that said, let's return to our focus on growing Biblical flowers for container gardens.

Which Biblical flowers to plant in whichever containers you elect to use is your choice. All Biblical flowers can be grown in containers for glorious blooming beauty on porch, patio, balcony, or windowsill. By providing the care they need, they will reward you splendidly. Details about how to plant and grow each one are included with the specific flowers themselves in Chapters 2 and 3, where you'll find descriptions of them and their cultural needs.

Consider those needs and provide them as closely as possible in containers with the water, plant food, and location that they require. No doubt you'll need to do some experimenting with conditions and situations. That's part of the fun of gardening and container gardening in particular. We can all be plant experimenters and explore new growing horizons with our families and to share lessons learned with friends and neighbors too. In the meantime, there is one special area of container gardening that allows you to focus on coaxing nature to do your bidding. Try forcing Biblical flowers to bloom indoors when winter winds wail and snow dusts your doorway.

WINDOWSILL BIBLE FLOWER GARDEN FOR CHILDREN

Youngsters, and indeed your entire family, will appreciate God's glorious world even more when they plant and cultivate their own beautiful Windowsill Bible Flower Garden. Better yet, this colorful discovery project enables children and families to truly grow together in faith and understanding as they enjoy God's beautiful flowers, wherever they live.

Coaxing Biblical spring-flowering bulbs to bloom during winter is easier than you think. Your youngsters and their friends can watch colorful crocuses, narcissi, hyacinths, and tulips bloom

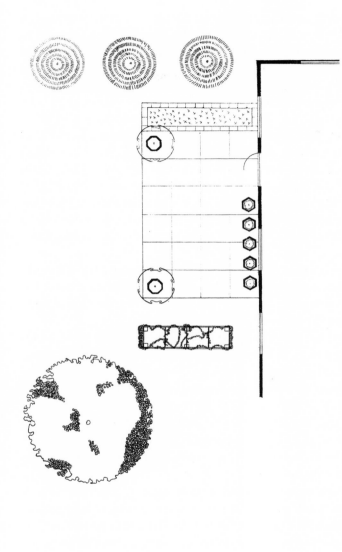

If you have limited backyard space, enjoy a patio planting of Biblical plants. In this sample design, Biblical vegetables and/or herbs can be grown in large tubs, or in a raised bed planter near the patio door for herb cooking pleasure. In larger tubs you may try dwarf fruit trees; add a small grape arbor, oak, poplar, or cedar tree as you wish to complete your Biblical plant décor.

indoors, even as the snow falls outside. Even better, as they grow and enjoy these Biblical flowers, they can look up references to them in the Scriptures to appreciate even more the wonders of God's world.

You can enjoy their beauty and special meaning even if you don't have any backyard garden space. Indoor windowsill bulb gardening is good, old-fashioned fun and virtually foolproof. The plants and flowers are preformed inside the bulb. With water and warmth, the bulbs do the work. You can usually find the desired, large-size bulbs that are best for indoor forcing in local garden centers. Mail-order catalogs also offer desirable big bulbs for forcing. You'll find a list of reputable mail-order firms, including some that specialize in bulbs and have a variety of Biblical flowers, in Chapters 11 and 12. All provide colorful free catalogs.

Timing is important to achieve the blooming period you desire. For flowering in January, plant bulbs around October 1. For February flowering, plant in mid-October. For March and April blooms, pot bulbs in mid-November.

You can use almost any kind of container for growing bulbs and forcing them to bloom indoors, but it must have a drainage hole or some provision to let excess water drain away. The container should be at least twice as high as the bulb to allow for adequate root development. Garden centers usually have bulb pans or pots just the right size. Tell the specialists there about your project and they can probably provide not only the best bulbs, but containers, soil mix, and necessary fertilizer for proper plant nutrition, too.

Flower bulbs prefer loose, crumbly soil. It is best to avoid outdoor garden soil, which may have unseen fungus and disease organisms in it. Best bet is to buy house plant potting soil mix. The alternative is to use any good-quality soil mix available at garden centers. Actually, you can use just a mixture of sand and peat moss, because all the food your plants require to bloom is already stored in the bulbs.

Potting is the easiest part of all. Ask your youngsters to place a small stone or piece of broken clay pot over the drainage hole in the bottom of the empty pot. This prevents clogging and keeps soil from flowing out. Pots 6 to 10 inches in diameter and 4 to 5 inches high are good for a grouping of bulbs. If you plan on using an indoor windowsill for display, made sure your pot will fit that spot. You can also use other types of more decorative containers. Just follow the tips here and you should enjoy blooming and fragrant success.

Fill the container with just enough of the soil or potting mixture so that when the bulbs are placed on top of this layer, the tops of the bulbs will just reach the rim of the container. You can plant as closely as you like, so they almost touch one another. For the fullest, most colorful and dramatic effect when they flower, use as many bulbs as will fit in the container or bulb pan neatly, with a half-inch space between them. Next, press bulbs gently into the soil mix. Don't force them down, as you might damage the base of the bulbs.

Once you have bulbs in place, fill the rest of the container with your planting mix. Be sure it settles between the bulbs. Pat the mix gently into place, but not too tightly. Then, cover the bulbs to within a half-inch of the top of the container. The tips of your bulbs should just peek above the soil. Finally, water the container thoroughly.

Be sure that you label each container with the variety name and color and the date the bulbs were planted. That gives you the names of those that performed best so you can focus on them in the future and recommend those varieties to gardening friends.

All spring-flowering bulbs must have forced "winter cold" or dormancy in order to develop their roots. In the northern part of the United States, spring-flowering bulbs go dormant in winter. Most Biblical flowers are bulbous—crocuses, daffodils, hyacinths, and tulips in particular. For forcing purposes indoors, you must provide a dormant period so that bulbs will be fooled into thinking they have had their resting period and it is now time for flowers to emerge and bloom, months before they really would in your outdoor garden in spring.

Most bulbs require at least thirteen to fifteen weeks of cooling. You can provide this necessary cold treatment by placing the containers indoors in a cold area, such as a northern corner of a basement or garage, or outdoors directly in the ground. Indoors, bulbs seem to root best in a cold, dark location under cover; an unheated garage, cold basement, or outdoor shed in which temperatures range between 40 and 45 degrees is best. Water these containers regularly. Another way to provide a period of dormancy is in the refrigerator. That will chill them to the necessary temperature and usually duplicates the dormant conditions close enough to give you the desired preparation for forced blooming.

Outdoors, you should cool the bulbs in a pit. Dig a bed or area that will hold all your containers. It should be an inch deeper than the tallest container. Set the pots close together and water them well. Then, cover them with 6 inches of soil or sand. That's all there is to that.

You can see the mound on the ground so you know where they are when it is time to dig them up to begin forcing their blooms indoors. Mark your calendar, so you can begin reviving your stored pots and bulb pans after the thirteen to sixteen weeks they need to set their roots. Check your stored pots, those indoors as well as outdoors, between the twelfth and thirteenth week. Bulbs are ready to be moved into the house when sprouts are well out of the bulb, about 2 to 3 inches high in most cases. When you bring containers indoors, place them in a cool room with a temperature between 55 to 60 degrees at first. Let them stay in that area, out of direct sunlight for two to three weeks. You'll notice that they will be growing faster during this period. When you see buds begin to form, you can bring them into their appointed blooming place, a windowsill, table, or focal point display spot. Never place any of these on a TV or against a hot southern exposure window. Remember, these are spring-blooming flowers accustomed to cooler spring weather outdoors. They don't need or want too much heat or strong sun. Continue to water your containers whenever they feel dry. Within three to four weeks of when you bring them indoors, your crocuses, hyacinths, narcissi, and tulips will be blazes of amazing color and beauty. While winter winds wail and snow drifts outside your home or apartment, your family, friends and neighbors, and their children will marvel at these glorious, colorful plants of the Bible on your windowsill.

Chapter Seven

❧

Cultivate God's Beauty Everywhere

Congratulations! You want to plant, cultivate, and enjoy a Biblical garden. Good for you and good for your family, friends, neighbors, and community. This chapter is designed as a collection of ideas from accomplished Biblical gardeners, which I hope imparts some of their feelings of faith and for the land. In this chapter, I've included ideas, pointers, and some basic designs that can be used or modified as desired. Equally important are the designs, ideas, tips, and advice shared by the founders, curators, and faithful cultivators of Biblical gardens.

There are opportunities for youth groups at Sunday schools or in programs at churches and temples as well as in religious schools to grow meaningful Biblical gardens. To help, I have worked with my son, Peter Swenson, who is a horticultural landscape graduate, to provide two basic Biblical Garden designs. These are based on our Judeo-Christian traditions that are rooted in the Holy Land. The Star of David and the Christian Cross can be changed in size based on land available. Choice of flowers is up to those planting their garden. He also designed other suggested gardens for this book. Feel free to use or adapt them as you wish.

CROSS AND STAR OF DAVID BIBLICAL FLOWER GARDENS

These two religious designs provide options for a variety of Biblical flowers. By selecting spring-flowering crocuses and narcissi, the designs will bloom early. Interplanting with hyacinths, anemones, and tulips, the blooming season is further extended. Adding irises and lilies will provide profusions of blooms into early summer and even through August. By selecting different varieties among narcissi, hyacinths, tulips, irises, and lilies, gardeners can stretch the blooming season for each type of flower. For example, some narcissi bloom very early, others midseason, and others later during spring. That fact also applies to irises, tulips, and lilies. Plant breeders have developed varieties that bloom early, mid, and late for that type of plant. Good for them and great for gardeners to extend glorious blooms over more weeks.

Here are basic Biblical flower plantings for these two designs. The original drawings were done on an 8½- by 11-inch-page with 1 inch representing 2 feet in the actual garden. In both designs, the letters represent different flowers.

STAR OF DAVID BIBLICAL GARDEN

For the Star of David design on the next page, A is Madonna lily with a total of 18 bulbs. B is hyacinth with 12 bulbs, or you could double or triple that number in their assigned location. C is for daffodil or trumpet narcissus and 36 are suggested. D is the area for massed crocus, either snow crocus or giant crocus and 250 bulbs fill the area. Saffron or fall crocus can be used or species tulips as a substitute and 225 bulbs fill the design area for them.

The interior bed lines are for demonstration purposes in this design and are not meant to represent rows of plants on ground. The center of the star can be used to display a statue, a birdbath, or other focal point, including a Biblical tree if desired. Species tulips such as *Tulipa bakeri, T. aucheciana,* or *T. tarda* are better choices, I believe, than hybrid tulips. All bulbs except lilies

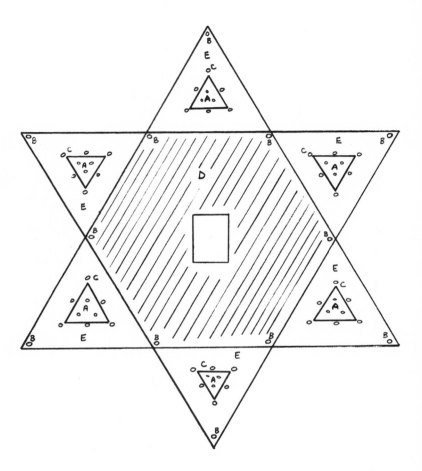

The Star of David offers its own distinctive religious design, part of the Judeo-Christian tradition. You can plant a variety of bulbs as you would for a Cross garden to have flowers for months. Crocuses and early daffodils begin the blooming season in March and April and you can then even plant lilies to give you bloom in July and August. See the list of flowers used for the Cross and adapt as you like.

should be planted 6 inches apart and the lilies 12 inches apart in this design.

PLANTING THE CROSS GARDEN

For the Christian cross planting bed (see page 166), the original design was drawn as 1 inch equal to 2 feet of garden ground. There are 3 Madonna lilies in the plan (A), 12 hyacinths (B), 36 daffodils or trumpet narcissus (C), and 104 crocus (D). Depending on the actual size of the ground, you would need to adjust the number of bulbs and corms for mass displays. Some gardeners prefer to see a few representative specimen plants while others prefer massed color displays. You can use these basic design thoughts as you create your garden, using whatever Biblical flowers you prefer. Growing with God is one project that allows you to expand and explore your own growing horizons. Other Biblical flowers can be substituted, such as iris and tulips or some of the other flowers in Chapter 3 that aren't necessarily mentioned in the Scriptures, but do have Holy Land roots.

If you have drawing and design capability on your computer, you can produce basic garden plans. There also are several excellent computer garden landscaping software packages available at Staples and Office Max and other computer stores. These let you do more elaborate garden designs and even see colorful graphic results of plants in bloom at their mature size.

As you plan your Biblical garden, keep these key points in mind. Do you want to grow flowers that most closely look like those mentioned in the Scriptures? If so, concentrate on the flowers in Chapter 2. Do you want to have a wider choice of flowers? If so, consider adding those in Chapter 3, which trace their roots to the Holy Land and have wild relatives still growing there. Do you want a dramatic flower garden with the biggest, most colorful blooms? If so, then you can consider flowers related to those in Chapter 3 that have been created from heirloom flowers by plant hybridizers. Modern hybrids that provide huge blooms and a longer blooming season may not look like flowers of the Holy Land but they do share the same flower roots.

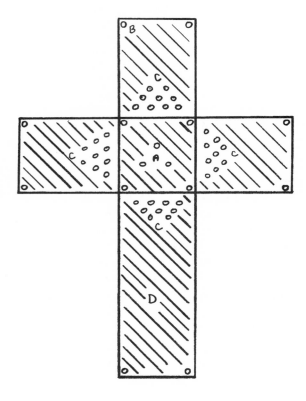

Extend the blooming season in the Christian Cross Biblical garden or any others by planting different flowers to provide early spring to fall blooming sequencing.

For March—Crocuses, and hybrid giant crocuses.

For April—Early-, mid-, and late-season daffodils.

For May—Early and late varieties of hyacinths.

For May and June—Early, mid, and late types of tulips.

For July—Try various types of iris for several weeks of bloom.

For July and August—Lilies of various types including Oriental and day lilies.

For August and September—Fall-blooming anemones and fall saffron crocuses.

Some gardeners may wish to specialize in just a few types of flowers, more closely planted to achieve dramatic masses of blooms in each flower's traditional season. Others may wish to use as many different flowers for a longer season and greater opportunity to achieve a wider range of Biblical plants. You may wish to copy these designs or draw your own patterns. Then, simply stake out your garden area and plant those flowers you wish to have. You can also add others in future years to extend the blooming season or vary colors, sizes, and shapes from among the other Biblical flowers.

In addition, there are other garden designs and plans based on Biblical gardens that have been successfully growing around the country. Many thanks to those who sent in plot plans, details, and sketches that are examples of what can be accomplished in smaller spaces and different places. Unfortunately, space considerations precluded using many in this book. You can visit some of these gardens via the Internet, see the plots and plants, and transplant some of their designs and ideas into your garden. (See the list of websites on page 257.)

SOLVE SLOPING LANDSCAPE PROBLEMS

Sloping areas present problems for landscaping, especially when too steep for lawns. In the Holy Land, many plants grace slopes and hillsides. You can use Biblical flowers on your sloping areas. Anemones, crocus, cyclamen, and narcissi all grace the land in their wild and wonderful ways. That natural ability makes them good choices for slopes. Rock gardens are another solution. Positioning rocks on slopes sets the stage for these delightful flowers to be placed among stones to give a similar look as one would see in their Holy Land rocky habitat.

Trees and shrubs are also useful for solving slope problems. Once established, their roots hold the soil and ground covers can be used beneath them. Some of the trees that trace their roots to the Scriptures are attractive additions. Hawthorn, flowering almond, and mulberry add color in spring and can become decorative accent trees. Oak, poplar, and willow are others mentioned in Scriptures that can be used, depending on the soil situation. Willows serve best in moist areas, but be certain never to plant them near septic systems or leach fields because their roots are invasive in search of water and can clog underground drainage pipes. Pin oaks do well in moist areas where other trees won't prosper.

Narcissi are ideal for naturalizing on slopes, under trees and shrubs, and in meadows or other areas allowed to grow more naturally than trimmed lawns. There are many different types of narcissi that are appropriate for naturalizing. Mail-order firms offer special collections of the varieties that do best in that habitat. Some varieties have more color and are fancier with longer trumpets, which provide an array of appealing blooms of different shapes. There are literally hundreds of different types of daffodils available today, packed into colorful pages of free catalogs available by writing to the lists of garden firms included in Chapters 11 and 12.

TRY GROWING WILD

Growing wild and wonderful has other advantages, besides saving lawn-mowing time and expense. One year I left a quarter-acre of lawn unmowed to see what would happen. First bluets, those first tiny floral harbingers of spring appeared, followed by small blue violets. By June hundreds of daisies were blooming. Then black-eyed Susans blossomed and by fall asters were plentiful. Obviously the wildflower seeds were still in the ground and viable from years ago when the land was originally a meadow.

Because we had not used any herbicides, the wildflowers emerged in their accustomed season when I stopped mowing. The result was a colorful meadow garden, with its share of tall grasses, but those, too, had appeal as they matured and even formed seed heads that swayed in the breeze. To that meadow I added several dozen narcissi in various places. As perennials, they bloomed every year and the wild lilies also added their summer reddish and orange colors.

When planning Biblical flower gardens, look out of your kitchen window and windows of other rooms you use frequently. A garden that can be seen from those windows will be a daily treat. From sighting the first crocus or daffodil to dramatic displays of lilies in summer while cooking or washing pots and pans are special joys from gardens that we can see from home windows.

CONSIDER MEDITATION GARDEN AREAS

Privacy gardens for prayer and meditation are becoming increasingly popular, judging from the many people I've interviewed for this book. Rev. Hudson-Knapp reports that he has seen people stand at the water garden area in deep thought and then quietly walk away, or sometimes come into the church. Joseph Scott also noted that people driving past their Biblical garden would stop their cars and visit the garden during its spring and summer blooming times.

For meditation or privacy, coniferous trees are useful. Cedars and pine trees can screen unsightly views and reduce noise. Research has revealed that planting trees along highways and city streets can greatly reduce noise levels and provide more peaceful, quieter garden sites. Shrubs can also screen views and soften sounds that abound in cities.

Look around your land for a possible meditation area. Biblical flowers add that extra-special meaning. A bird bath, a bench, some special statuary can be a decorative accent to that special spot for you and others who wish to find a quiet place for privacy. Water gardens have been rediscovered for the soothing effects that bubbling, running water provides. That has been understood for years in Japan, where water gardens with lilies and fish are popular. At last that wonder of the natural world is being better understood and included in garden plantscapes in America. In fact, one of the fastest new trends in gardening is the addition of water gardens with fountains, streams, and small ponds.

BIBLICAL GARDENS ARE THERAPEUTIC

Biblical flowers have that extra dimension and meaning from Scriptures and their roots in the Holy Land. Charles Sourby has been growing Biblical flowers for their therapeutic values among older and ill people at nursing homes. He discovered in his work at hospitals, nursing homes, and elder-care facilities that Biblical flowers actually helped patients recover health and regain their sense of perspective. In fact, in this decade there has been renewed attention paid to the value of gardening, especially the fragrance of flowers and herbs to revitalize people who had almost lost interest in life. If you have older members in your family who have illness, consider how some Biblical plants and visits to a Biblical garden can lift their spirits.

Congregate housing facilities have also begun to recognize the healing value of plants. I verified that fact at various senior care homes where I have been giving my garden slide programs for the past few years. A majority of residents of these senior centers and nursing homes I have met are people of strong faith. Nurses have told me that having a potted Biblical flower to tend has brought new interest to older citizens, whose hearing, sight, and senses had been dulled by age, according to one of my sources. The scent of a hyacinth seemed to reawaken something important and give patients a needed boost. Perhaps you may consider planting some Biblical flowers for residents of senior housing facilities in your town.

With these thoughts about Biblical garden planning and the suggested designs in this chapter to guide you, it is time to hear some observations and good advice from veteran Biblical gardeners I've interviewed for this book. You'll also find more information about these gardens in Chapter 9, "Biblical Gardens to Visit in the United States."

PRINCE OF PEACE BIBLICAL GARDEN

Joseph Scott, a parishioner of Prince of Peace Lutheran Church in Augusta, Maine, had an idea for a Biblical garden after retiring. He approached his pastor, Rev. David, about the idea and then the church board. After the entire congregation gave its hearty approval, Scott began his 20- by 24-foot garden right on the church's front lawn. He correctly foresaw that the blooming beauty of the garden would attract passersby.

Scott prepared and enriched the soil, included rocks and a ledge to mimic the hilly topography of Palestine, and then began planting. Today more than 40 different plants grow there. The flowers it includes are numbered for easy identification in a handy booklet, corresponding to numbers on stakes: poppy, mallow, flax, anemone, iris, chamomile, sage, crown daisy, lily, crocus, narcissus, tulip, cyclamen, and rose.

"This has been a real group effort," Scott points out. "The garden includes a variety of forty-seven different plants native to the Holy Land. Some of the plants are substitutes because the species or variety are difficult to obtain. When a substitute is made, the same genus and family is used and looks very much like the one found in the Holy Land," Scott explains. "The garden is designed so if you cut a chunk out of a field in Israel, this is what you might see."

Its garden is a valuable asset to the Prince of Peace Lutheran Church in Augusta, Maine, the state capital, attracting viewers and new visitors to the church, who stop to see the garden. "It is a great educational tool and also has brought many different churches together for their common interest, the Bible," Scott points out.

By way of growing together, Scott notes, Rabbi Susan Bulba Carvutto brought the congregation of Temple Beth El of Augusta, Maine, to see the garden. "They were very enthused and thought it was a great idea," Scott recalls. He estimates between twenty-five to thirty inquirers visit the garden weekly. A twenty-page booklet describing the plants is available in a weather-protected rack at the garden and includes appropriate Scriptural quotations.

As a dedicated Biblical gardener, Joseph Scott remains very busy, sharing his passion for these flowers and plants and their significance to all people of faith. He regularly conducts slide programs of the Prince of Peace gardens that also include many other dramatic color slides he has collected through the years while on the staff of the New York Botanical Gardens and as Maine State Horticulturist.

SHIR AMI BIBLICAL GARDENS

The Shir Ami Biblical Garden is thriving, carefully tended by Jules Hyman and Stan Averbach of the Shir Ami Bucks County Jewish Congregation, located in Newtown, Pennsylvania.

The concept of the Biblical garden began in 1999 during Bible study when the synagogue decided to use grape juice in lieu of wine for sacramental purpose since children were frequently involved. Jules remarked that they should construct an arbor and make their own wine. Although tongue in cheek, Jules had an arbor constructed with the blessing of associate Rabbi Shira Joseph. Stan Averbach was named horticulturist to select appropriate grapes for the empty plot of ground. It was decided to share the tasks, Jules to handle administration and Stan to tend horticultural aspects.

One key factor was how to finance the gardens without becoming a burden to the synagogue. Together the cochairmen organized a variety of fund-raising projects focused on dedication of planters, garden pathway stones, and plants themselves. Congregation members bought plants in honor of relatives. Other bought pathway stones, and containers for plants. A children's garden was an important part of the plan to appeal to young members. It is adjacent to the synagogue school and features flowers, vegetables, and herbs planted in Earth Box planters. Each planter has been dedicated. The walkway itself has room for 180 round pavers.

"We make an impression of a child's hand, their name, and birth date. Early on we completed 144 for a contribution of $37 each. We are very proud of our congregants for this support. All dedications are in perpetuity. If a plant dies, regardless of size, it will be replaced at no charge. Jules and I believe that one secret of the garden's success is that our congregants want 'Living' dedications that grow and flourish with time," Stan says. "Each school class will be assigned various containers. The purpose is to teach the responsibility for living things, that all living things, no matter the size or beauty, have a purpose in this world. They should be appreciated and respected. All plants have dedicated markers and a pamphlet serves as a guide for visitors. Docents act as guides for groups and the gardens are open to all without charge, but voluntary contributions are accepted and applied to the maintenance fund. The memorial and honorarial contributions to the gardens and those for other special occasions have been substantial. Every plant has or will have a marker with the name of the plant, its Latin name, its Hebrew name, the applicable passage in the Scriptures, both in Hebrew and in English, the dedication, and name of the dedicator.

"The next project is to have a committee large enough to help maintain this Holy place for the congregation and the entire surrounding community," Stan concludes.

MAGNOLIA PLANTATION BIBLICAL GARDEN, CHARLESTON, SOUTH CAROLINA

Since its inception, the Biblical garden has been a constant source of interest and enjoyment for visitors to it at the famed Magnolia Plantation in Charleston, South Carolina, according to Taylor Drayton Nelson. Thousands of families have visited it.

"We regularly receive calls from people requesting information about the Biblical garden, its origins, its plant material, and other details," Taylor Nelson says. As the grandson of the present owner, J. Drayton Hastie, he is following in the footsteps of a man who has given America what Charles Kuralt described as his "greatest Charleston pleasure."

"Often callers tell us that they are planning a Biblical garden for their church or community," Taylor Drayton Nelson wrote to me recently for this book. "As you know, the idea of creating a 'Biblical garden' is a natural to those interested in both gardens and religion. In fact, it makes sense to anyone interested in either, as the theme of the garden in the Bible is so overwhelmingly significant that it immediately brings the garden observer into contact with an 'idea' of the garden. The garden takes on a meaning beyond the simple arrangement of plant material," he explains. "That, it seems, is the aim of the best garden designs. In the Biblical garden, the framework for the actual design has been in the collective consciousness for two thousand years."

With grateful thanks to Taylor Drayton Nelson and his grandfather, J. Drayton Hastie, I include here their helpful tip. "As far as sources for plant material are concerned, Forest Farm Nursery, 990 Tetherow Road, Williams, Oregon, 97544-9599, has a good selection of rare and hard-to-find plants, some of which are appropriate for Biblical gardens," Taylor adds. With extra thanks, I have included them with other key sources in Chapter 11.

TEMPLE BETH SHALOM, SUN CITY, ARIZONA

Although Hy Mandell, chairman of the Biblical Garden at Temple Beth Shalom, calls their gardens modest, they are truly impressive, especially because they are in a desert climate. The garden area is 138 by 80 feet in the main section and 40 by 50 feet in an adjoining section. All trees and plants are identified by individual markers with Scriptural reference in Hebrew and English. They are constantly escorting church and other groups through the garden and have a brochure that is helpful in describing and identifying the plants. Identifying plants and including their Scriptural references is a useful idea for every Biblical garden. It is something we all should consider, especially if our garden will be at our house of worship and open to the public. Simple brochures about the plants and temple or church also prove welcoming greeting cards for those who stop to visit the garden. This Biblical garden seems to beckon to visitors, according to Hy Mandell, "Temperature permitting, the garden has been a magnet for meditation."

RODEF SHALOM BIBLICAL BOTANICAL GARDEN, PITTSBURGH, PENNSYLVANIA

One of the oldest and most extensive displays of Biblical plants in North America, this garden has been designed to depict an area reminiscent of Israel. It has a small desert section, a stream reminiscent of the Jordan and springs that should evoke En gedi in the Holy Land. Because a high percentage of the plants are tropical or semitropical, they are brought in each fall and maintained through the winter.

Each summer this temple conducts special exhibits and programs about Biblical plants and history of the Holy Land. For example, one focused on King Tut's Vegetable Garden, another on Fragrance Through the Ages—Perfume, Incense, and Cosmetics. These gardens feature more than 100 plants in a setting of cascading waterfall, desert, and river scenes. All plants are labeled and a Biblical verse accompanies each plant. Tours through the garden are led by trained docents. In addition, there are hosts and hostesses who assist with thousands of visitors who come to that garden each summer.

FIRST CONGREGATIONAL CHURCH OF FAIR HAVEN, FAIR HAVEN, VERMONT

Rev. Marsh Hudson-Knapp has provided details about some of the numerous projects that use the Biblical gardens to enrich their communal and individual spiritual lives. Your church, synagogue, or home garden might adapt some of them for your own ministries.

A prayer guide is available to the children's Biblical vegetable garden, a circular area with a park bench nearby to sit, observe, read, and talk with children. In the garden itself, many plants are included; lentils, hyssop, coriander, cantaloupe, and broad beans grow as well as dill, chickpeas, barley, and myrtle. Among flowers, children and parents will see chamomile, violets, and poppies. The accompanying prayer guide invites visitors to sit down and enjoy their time with God and the divine creation.

It concludes with this important message, an imaginative prayer dialogue between visitor, the elements of the garden, and God. "Teach us, loving God, to cherish and tenderly care for your earth! I am the water that brings life to dry seeds, that quenches the thirst of growing plants. God sprinkles me from the skies like holy water to bless earth. You do not command when I will come. God calls me forth and sends me out. You need me. You need God. Never forget this.

"We are the plants and elements of the Garden of God. Learn how to live in harmony with us, and we will bless you. In every plant that grows in the earth and the waters of life, you speak to us O God. Teach us to listen and love with our eyes and noses, with tongues and bare toes, with our hearts and open souls, Amen!"

As Rev. Hudson-Knapp tends his garden and guides his congregation, he adds several other points about the use of Biblical flowers and plants in ongoing ways, in services as well as around the garden.

"Our church tries to include flowers from the Biblical gardens at public dinners, on the communion table for worship, and in every aspect of our church's life. We include news about the plants in bloom in the church bulletin with Biblical background about them. Plants also are illustrations in sermons and the gardens are a place for Sunday fellowship time and daily gatherings of folks from town. There are tales about Biblical plants to enliven children's story times and we also involve children in planting and caring for the garden, drawing Biblical plants, and then actually picking and eating the plants and taking some home. We also serve Biblical fruits and vegetables as available at our public dinners," Rev. Hudson-Knapp says. More details about this attractive garden are in Chapter 9, "Biblical Gardens to Visit in the United States."

ST. GREGORY'S EPISCOPAL CHURCH, LONG BEACH, CALIFORNIA

Betty Clement had envisioned a Biblical garden after her purchase of Eleanor King's and Allan Swenson's Biblical garden books. Her dedication saw it come into being. "When Edith Place Vander Meulen moved to Long Beach we asked permission to create a Bible garden," Betty Clement recalls. "With Edith as the chair and myself as cochair and researcher, we set out to raise necessary funds. The ECW women of the church and the vestry as well as individuals were generous. Church friends cleared the central courtyard of existing plants. A local landscaper removed some brick, created crushed rock paths, improved the soil in four planting areas. He also built a piled rock fountain and provided our first four potted trees, a braided ficus representing the Trinity, a date palm, an olive, and a pomegranate," she noted.

Visits to local nurseries, library research, contact with the Cathedral of St. John the Divine in New York, and the Los Angeles Arboretum helped locate plants. By 1991 they had acquired 56 plants. Their search continued. Today, their list includes 86 Biblical plants, including some flowers that few other gardens have been able to locate, proving that faith and good old-fashioned perseverance do lead to accomplishments.

Inside the church the Biblical garden focus also led to flowers being depicted on kneeler cushions for use inside St. Gregory's main altar rail. During the 1990s, thirteen women created six sedilia cushions, five kneelers for the prie-dieux, and a hassock for one sedile. They were designed by Lee Stanley and depict plants of the Bible. Seven of the plants mentioned in the Bible were depicted on the new cushions. These include the red anemone, pink cyclamen, and yellow sternbergia, all "flowers of the field" which nourish the spirit. The barley and wheat, grapes and pomegranates, which nourish body and soul, also have been faithfully done in needlework on cushions. A Needlepoint Dedication was held on Trinity Sunday, May 25, 1997.

"Creating a Biblical garden has become a long-term endeavor, but it is one that will bring many years of learning and pleasure to our parish and, hopefully, to our community," Betty Clement believes. "It takes a church to grow a garden," Betty Clement explains. She has high praise for her friends at St. Gregory's. Shirley de Ment is the church "artist in residence" and has done exquisite line drawings of many plants. Others active in garden projects include Bea Evans, Christopher Johns, Lois Chaddock, Harriett and Les Lundeen, Ella Mae Marble, Gina Locke, and Rose Nelson.

COMMUNITY CHURCH BIBLICAL GARDEN, FAIRFIELD GLADE, TENNESSEE

Grace Westlake thoughtfully described some of the features of the Biblical garden she helped plan and tend at Community Church in Tennessee, affiliated with the Presbyterian and United Church of Christ. It was the first church that had sufficient room to develop a garden, she notes. On the Cumberland Plateau area, the selected spot had trees, grass, and lots of rocks, but offered a 16- by 70-foot garden area between two sections of parking lot.

Once the soil was dug, turned, and ready, the design took several appealing forms. A wheel area was set up with pie-shaped beds for various herbs. Adjacent to the wheel bed a triad garden of seven beds was laid out. It included saffron crocus, iris pseudocorus, and hyacinths. The triad, an equilateral triangle, has long been considered a symbol of the Trinity, Grace Westlake explains.

The cross at the garden forms four beds. At the top is a globe thistle in front of loosestrife. On the right is mallow. The bottom bed includes flax. At the four corners of the cross Madonna lilies were planted and, at the tip of the cross, anemones. The cross, the focal point of that beautiful garden, is one of the oldest and most universal symbols for Biblical gardens. The Community Church gardens also include a ladder design garden, a plant legend garden, and various trees and shrubs around the area along with grape vines. All these designs can be readily adopted for a variety of Biblical flowers, depending on preference and the time required to plant and tend them.

BIBLICAL GARDENS WEBSITE HAS VALUABLE TIPS

One especially useful website is Biblical Gardens at www.Biblical Gardens.com. Shirley Sidell is the founder and website gardener there. Several years ago she announced her vision to encourage a Bible garden in every home, church, synagogue, or temple using the Internet to exchange ideas and knowledge. In the process, she has traveled overseas and around America obtaining a wealth of knowledge which she shares on her website. That website is made up of people from around the world who have an interest in the research and application of Bible gardens and gladly share their enthusiasm, knowledge, and contacts with all visitors.

A necessary part of her project was to build a historical database of plants mentioned in the Bible. It is there to tap today for both members and nonmembers. There also is a Kids' Corner and links to obtain kits of Biblical plant seeds too. From that site you can link with other Biblical gardens to see how they have done their plantings and harvest some of their information too. Free seeds are offered by Internet and the Biblical Gardens Free Seed Offer, P.O. Box 31342, Walnut Creek, CA 94598. According to Shirley Sidell, all seeds have been home grown by volunteers and lovingly hand gathered. She also offers Bible garden kits—a Bitter Herb Kit, a Wild Flower Mix-Cut Flower Mix, and a Biblical Garden Kit. That is a collection of plants mentioned in the Bible or substitutions that are in the same families or similar to the species grown in the Hold Land. The package contains at least 25 seeds of 20 different plants, many hard to find, Shirley explains. Planting directions and a booklet of Biblical references are included in the pack. The seeds may include Shasta daisies, salvias, poppies, mallows, hollyhocks, altheas, and others.

This notable site and other worthwhile Biblical garden websites are included in Chapters 11 and 12. They provide exceptional additional resources for all future Biblical gardeners.

Beginning in 2002, I'll also be providing more good growing information as part of an expanding national website, *www.norellsoftware.com.* You are invited to tap into what will be an ever-growing source of gardening information, plus cross-referenced sources for plants all around America. If you have Biblical gardening, or general gardening information to share, please e-mail it to me at aswenson@gwi.net. If you have photos or slides to share, please send them in JPEG format. My snail mail address is:

> Allan A. Swenson
> Windrows Farm
> 920 Alewive Road
> Kennebunk, ME 04043

Chapter Eight

☙

Flowers for Holy Days and Holidays

Each holiday is marked by cherished traditions that bring joy, comfort, and warmth and provide continuity from one generation to the next. Flowers are as much a part of people's lives throughout the aeons of recorded history as anything else you can name. Some flowers are traditionally associated with the major holidays, especially Christmas and Easter, and have been for as long as we can trace the records. In this chapter you'll learn more about the flowers that are most closely related to our major religious holidays with tips to tend them better.

EASTER LILIES

For many, the beautiful trumpet-shaped white flowers of the lily, *Lilium candidum*, symbolizes purity, virtue, innocence, hope, and life, which encompass the spiritual essence of Easter.

History, mythology, and the world of art are filled with stories and images that speak of the beauty and majesty of the elegant white flowers. Dating back to Biblical lore, the lily is mentioned several times in the Bible, one of the most famous references being in a description of the Sermon on the Mount. Here Christ tells his listeners: "Consider the lilies of the field, how they grow: they toil not, neither do they spin; and yet Solomon in all his glory was not arrayed like one of these."

Easter lilies have often been called the "white-robed apostles of Hope" by religious leaders. Christian tradition also says that lilies were found growing in the Garden of Gethsemane after Christ's agony. Another tradition has it that beautiful white lilies sprang up where drops of Christ's sweat fell to the ground in his final hours of sorrow and suffering on the cross. Churches continue this tradition at Easter time by banking their altars and surrounding their crosses with masses of Easter lilies, symbolizing the resurrection of Jesus Christ and hope of life everlasting.

Easter Lilies Have a Long Tradition

Lilies have played significant roles in allegorical tales concerning the sacrament of motherhood as well. The pure white lily has long been closely associated with the Virgin Mary. In fact, no artist seems to have lived in the Middle Ages who has not painted a picture of the Blessed Virgin with a white lily. Fra Angelico, Titian, Murillo, Botticelli, Correggio, and others captured the beauty of the white Madonna lily. Actually, in the year 1618 a special papal edict laid down stringent rules "as to the proper treatment of certain sacred subjects in art and the necessity of the introduction of the white lily into pictures treating of the Immaculate Conception. In that edict, it also was commanded that roses, palms, and lilies should be employed as the flowers suitable to be scattered by angels or as the decoration for paintings, according to Harold Moldenke from his careful Biblical plant research.

One legend is told that when the Virgin Mary's tomb was visited three days after her burial, it was found empty save for bunches of majestic white lilies. Early writers and artists made the lily the emblem of the Annunciation, the Resurrection of the Virgin: the pure white petals signifying her spotless body and the golden anthers her soul glowing with heavenly light.

In the thirteenth century, writer Bartholomeus Anglicus wrote, "The lily is an herb with a white flower and though the leaves of the flower be white, yet within shines the likeness of gold." As the saying goes, "To gild a lily is to attempt foolishly to improve on perfection." In yet still another expression of womanhood, lilies had a significant presence in the paradise of Adam and Eve. Tradition has it that when Eve left the Garden of Eden, she shed tears of repentance and from those remorseful tears sprang up lilies. The spiritual principle here is that true repentance is the beginning of beauty.

The white lily is a fitting symbol of the greater meaning of Easter, a mark of purity and grace throughout the ages. In millions of churches and homes, the flowers embody joy, hope, and life. Whether given as a gift or enjoyed in your own home, the Easter lily serves as a beautiful reminder that Easter is a time for rejoicing and celebrating.

However you wish to consider the botanical, scriptural, or historical factors surrounding the white lily, the Madonna lily as the Easter lily, they remain today deeply rooted in our Christian traditions. These fragrant white flowers make a welcome and meaningful gift that embodies the very essence of the celebration of Easter. Whether you receive one or plan to give the potted plants as a gift, here are important tips to help make your Easter lilies keep on giving their beauty longer.

Pointers for Choosing Lilies

Two of the greatest charms of the Easter lily as we know it today are form and fragrance. Look for high-quality plants that are healthy, have no insect or disease damage, are well potted, and are aesthetically pleasing from all angles. Select medium to compact plants that are well balanced and proportional in size; not too tall or too short.

One special point to achieve the longest possible period of enjoyment in your home is to find plants with flowers in various stages of bloom or "ripeness." This means the best selection would be a plant with only one or two open or partially open blooms and three or more puffy, unopened buds of different sizes. The ripe puffy buds will open up within a few days while

the tighter ones will bloom over the next several days and perhaps extend the blooming period beyond that. As flowers mature, remove the yellow anthers before pollen starts to shed,
especially if anyone in your household has an allergy. This also
gives longer flower life and prevents the pollen from tainting the
white flowers. When a mature flower begins to wither after its
prime, simply snip it off to make the plant more attractive as
other newly opened blooms appear.

Another point about picking plants. Check them out for an
abundance of dark, rich green foliage that is attractive and also a
vital sign of good plant health. Foliage should appear dense and
plentiful, all the way down to the soil line, which is a good indicator of an active, healthy root system.

Beware of Easter lilies displayed in paper, plastic, or mesh
sleeves. These should be used solely for shipping and removed
immediately upon arrival at florists or stores. Quality of plants
will deteriorate if they are left in sleeves too long. You should
also avoid waterlogged plants, especially if the plant looks
wilted. This could be a sign of root rot. We Americans tend to
overdo things, and overwatering plants clogs the soil pores and
rots roots and is the biggest killer of house plants.

Tips for Easter Lily Care

Easter lilies prefer moderately cool temperatures in the home.
Recommended daytime temperatures are 60 to 65 degrees F,
with slightly cooler night temperatures. Avoid placing plants
near bright, sunny windows where they will become overheated.
Also avoid placing them near drafts and exposure to excess heat
or dry air from appliances, fireplaces, or heating ducts. A lily will
thrive near a window in bright, indirect natural daylight, but
avoid glaring direct sunlight of southern exposure windows.

Lilies prefer moderately moist, well-drained soil. Water the
plant thoroughly when soil surface feels dry to a light touch, but
avoid overwatering. Use the toothpick test by pushing a wooden
toothpick into the plant soil. If it has soil particles clinging to it,
wait a day to water. If dry, give it a watering. If the pot is still

wrapped in decorative foil, be careful not to let the plant sit in trapped, standing water. For best results, remove the plant from decorative pots or covers, take it over the sink, and water thoroughly until water seeps out of the pot's drain holes. This will completely saturate the soil. Allow the plant to drain for a few minutes and discard the excess water before replacing it back into its decorative pot cover.

The growing pot is usually plastic from the producer, but also may be clay, which I much prefer because those pots drain and breathe better. After the last bloom has withered and been cut away, you can continue to grow your Easter lilies and even plant them outside in your garden to enjoy for years to come. Once lilies have finished flowering, place the potted plants in a sunny location. Continue to water thoroughly as needed, and add 1 teaspoon of slow-release 12-6-12 or similar analysis fertilizer every six weeks. If you prefer, you may keep your lily inside on a windowsill. However, you can move the pots to a sunny location outdoors after the danger of frost has passed.

Lilies Can Grow Outdoors, Too

To plant your Easter lilies outdoors, prepare a well-drained garden bed in a sunny location with rich, organic matter. Your compost pile can provide a good mixture. Use a well-drained planting mix or a mix of 12-part soil, 12-part peat moss, and 1-part perlite. I prefer to make mine a 4-part mixture with the fourth a good rich compost.

Good drainage is a key for success with lilies. To ensure adequate drainage, you may consider raising the garden bed by adding good topsoil enriched with compost to the top of the bed, thus obtaining deeper topsoil and a higher lily planting area. Carefully tap the pot and remove the bulbs with sufficient soil mix around them to plant in the ground.

Plant Easter lily bulbs 3 inches below ground level and mound up an additional 3 inches of topsoil over the bulbs. Plant your bulbs at least 12 to 18 inches apart in a hole sufficiently deep so that bulbs can be placed in it with the roots spread out and down as they naturally grow. Carefully spread roots and work the prepared soil in around the bulbs and roots, packing firmly to leave no air pockets. Water immediately and thoroughly after planting. Try not to allow the soil to heave or shift after planting.

As the original plants begin to die back, cut the stems to the soil surface. New growth will soon emerge. The Easter lilies, which were forced to bloom in time for Easter, under controlled greenhouse conditions in March, bloom naturally in the summer outdoors. You may be rewarded with a second bloom the first summer but most likely you will have to wait until next June or July to see these Easter lilies bloom again. They are perennials, so they should endure and bloom for many years.

Another important planting tip to consider is that lilies like their roots in shade and their heads or flowers in the sun. If possible pick a location where the sun reaches the growing lily plants, but the soil is shaded. Mulching helps conserve moisture in between waterings, keeps the soil cool and loose, and provides a fluffy, nutritious medium for the roots.

A more attractive alternative, of course, would be to plant a living mulch or low ground cover of shallow-rooted perennials. I prefer vinca or ground myrtle. Although not essentially a Biblical plant, some Biblical gardens do include them and they have their advantages, covering the ground and providing mass blooms in season. Stately Easter lilies rise above such low-growing plants. Both provide blooms and are reliable, sound gardening practices.

Easter lily bulbs are also surprisingly hardy even in cold climates. Be sure to provide winter protection by mulching ground with a generous layer of straw, leaves, or other materials. Come spring, carefully remove the extra winter cover mulch to allow new shoots to come up well. With these steps, your Easter lilies will continue rewarding you with beauty, grace, and fragrance for years to come.

The Easter lily capital of America is along a few miles of the Pacific Coast at the Oregon and California border. There an ideal combination of climate, soil, water, and dedicated people have developed this flower of such deep meaning, beauty, and tradition. The area offers a climate of year-round mild temperatures afforded by a protective bay, deep, rich alluvial soils, and abundant rainfall, the exact conditions needed to produce a consistently high-quality bulb crop. That accounts for the fact that more than 95 percent of the world's potted Easter lilies originate from this area.

Historically, in the late 1800s the Japanese grew most of these lilies, but World War II stopped imports. Current U.S. production began with a World War I soldier, Louis Houghton, who brought a suitcase full of hybrid lily bulbs to the southern coast of Oregon in 1919. He freely distributed bulbs to his friends and neighbors. When imports stopped, the value of lily bulbs skyrocketed. Many growing them as a hobby went into business. Easter lily bulbs at that time were called "White Gold," and by 1945 more than one thousand growers were producing them. Today, just ten farms in a small coastal region straddling the Oregon-California border grow the bulbs. Precise growing conditions are necessary since Easter lily bulbs must be cultivated in the fields for three to four years before they are ready to be shipped to commercial greenhouse growers for producing the Easter season lily plants.

Though the centuries, these lovely flowers have become a symbol of resurrection. Easter lilies rise from earthy graves as scaly bulbs to bloom into the majestic, fragrant flowers that embody the beauty, grace, and tranquility of that special season.

POINSETTIAS CELEBRATE CHRISTMAS TRADITIONS

Poinsettias have enjoyed popularity for decades, but they most certainly do not have their roots in Scripture or the Holy Land. The Poinsettia, often referred to as the "Christmas Plant," was first brought to America from its native Mexico in 1825 by the first U.S. ambassador, Joel Robert Poinsett. Hence its name. He was a Southern plantation owner and botanist and was struck by the beauty of the brilliant red plants he found blooming there during December.

Ambassador Poinsett had some of the flowers sent to his plantation in Greenville, South Carolina, where they flourished in his greenhouse. Their botanical name, *Euphorbia pulcherrima*, was given by a German taxonomist in 1833, but the common name, poinsettia, has remained the accepted name in English-speaking countries. The plant lives on and has become a living symbol of Christmas. Although not a plant of the Holy Land, it deserves recognition in this book because it has become so widely popular. More than 70 million plants are sold each year, making it the number one flowering potted plant in America.

Today, available in a multitude of colors, sizes, and shapes, poinsettias offer an abundance of holiday cheer for every décor. In tropical areas some grow almost to small tree size. We saw many during vacations and writing assignments on Caribbean islands, where they grow profusly in the wild.

The colorful parts of the poinsettia, commonly thought of as the flowers, are actually modified leaves called bracts. The true flowers are the small yellow "berries," termed cyantia, in the center of the bracts. As natural day length decreases in November and December, the bract color changes. While red poinsettias remain the overwhelming consumer choice, many novelty colors are gaining in popularity for holiday decorating and gift giving. Besides the red tones, now there also are white, pink, peach, yellow, marbled, and speckled poinsettias, too.

In recent years, plant breeders have hybridized exceptional new, bright, darker reds, to pinks and other shades, including marbled varieties. The first poinsettia most holiday shoppers are likely to see in stores is Eckespoint Freedom. Not strictly a new variety, Freedom has swept the poinsettia world since introduction only a few years ago. Freedom has become the top-selling poinsettia in the world today. Its ease of growth, earlier bloom, and deep ruby bracts against dark green leaves make it useful even for a Thanksgiving centerpiece. It also tolerates lower light levels and the less-than-perfect care typical in many homes. Freedom also is available in white, pink, and marble.

How to Pick the Best Poinsettias

When purchasing a plant to give as a gift, for church or your own home, keep these points in mind. Look for plants will fully mature, thoroughly colored and expanded bracts. Avoid plants with too much green around the bract edges. Unlike a rosebud, a poinsettia will not continue to mature in the home and may never achieve its full blooming potential if taken home prematurely, according to professional growers.

A good method of judging a poinsettia's maturity is making sure that the smaller bracts surrounding the cyantia are fully colored and lying horizontal. If these secondary bracts are not fully colored, the plant will quickly lose its color in the home environment. Make sure that the plant you choose still has the cyantia on it. Also, an abundance of dark, rich green foliage is a vital sign of good plant health and an active root system. Look for plants with plentiful foliage all the way down the stem. The leaves should be undamaged and free of discoloration.

If you purchase a poinsettia from a store that has kept the plastic or paper sleeve packaging on the plant, remember that if this has been on the plant for a lengthy amount of time, leaves may turn yellow and drop before the holiday season is over. As discussed, be wary of plants displayed in paper, plastic, or mesh sleeves. A poinsettia needs its space. The longer it remains sleeved, the more plant quality deteriorates. Sleeves are useful to protect plants during transport, but should be removed immediately upon arrival at the store.

Examine the soil of the plant. It is best to avoid waterlogged soil, especially if the plant appears wilted, which could be a sign of irreversible root rot. As I say many times in talks to garden groups, church organizations, and on TV programs, we Americans have a terrible tendency to overwater plants. That kills millions a year with misplaced "kindness" and is easy to avoid. When you shop for any plants, be aware and wary of any water-logged specimens.

More Poinsettia Pointers

Proper proportion of plant height and shape relative to pot size is another key to an aesthetically pleasing poinsettia. Plants should appear balanced, full, and attractive from all angles. A generally accepted standard is that the plant should be approximately 2½ times larger than the pot size. An 18-inch tall plant in a 4-inch pot will appear leggy, but the same plant in a 6-inch pot will look properly proportioned. In a 4-inch pot, heights under 12 inches look best.

For a beautiful presentation that lasts throughout the holiday season, it is important to select durable plants. The longest period of enjoyment is provided by plants with stiff stems, good bracts and leaf retention, and no signs of wilting, breaking, or drooping. Look for a care tag with detailed information on watering, light, and temperature requirements. If one isn't attached, ask for a care tag. Poinsettias are temperature sensitive, especially when you walk out of the store to take them home in the winter season, often with chill winds blowing. Just a few seconds of exposure to freezing weather may destroy a beautiful plant. Always protect plants from chilling winds and temperatures below 50 degrees F. Reinserting the poinsettia into a sleeve or large, roomy shopping bag will usually provide adequate protection for getting your plant home. Never leave your poinsettia in the car while you finish shopping. Pick it up as your last item and take it straight home. Immediately unwrap your poinsettia when you arrive at your destination, home, church, or friend's house.

In addition to a variety of red tones and other colors, you'll find poinsettias in new, exciting plant forms such as individual self-watering minipoinsettias. Plant producers are now growing these plants in special easy-care containers including those with self-watering systems, centerpiece and hanging basket types.

Larger forms such as massive color tubs and sculptured topiary tree forms work well in large offices and reception areas.

Care Tips for Poinsettias

The poinsettia thrives on indirect, natural daylight. Exposure to at least six hours daily is recommended. Avoid locations where the plant is exposed to direct sunlight as they may fade the bract color. If direct sun cannot be avoided, diffuse with a light shade or sheer curtain.

To prolong the bright color of the poinsettia bracts, daytime temperatures should not exceed 70 degrees F, according to Thom David of the Paul Eke Poinsettia Ranch in California. Avoid placing plants near drafts, excess heat, or dry air from appliances, ventilating ducts, or fireplaces. The color of your poinsettia will last longer with temperatures not over 75 degrees during the day and 60 to 65 degrees at night, according to specialists with Fischer Poinsettias. They also emphasize that you should be sure your plant never, repeat *never*, dries out.

It is not necessary to fertilize the poinsettia when it is in bloom. However, a balanced, all-purpose houseplant fertilizer may help maintain rich green foliage and promote new growth after the season. Always follow directions for the brand of fertilizer you use.

Since poinsettias are sensitive to cold weather, frost, and rain, outside placement during winter months must be avoided. However, in mild climates, an enclosed patio or entryway may be suitable, provided night temperatures never drop below 55 degrees. Make certain delicate bracts are protected from wind and cold rain. Chill damage will occur if temperatures drop below 50 degrees and will result in premature leaf fall. Exposure to frost, even for a brief time, usually results in the plant dying.

One of the most often asked questions about poinsettias is how to bring them back to bloom after the holiday period. If you are an enthusiastic gardener, you may wish to "reflower" your poinsettia for next year's holiday season. Remember, if you lose interest along the way, many others have abandoned the effort also. It requires a lot of dedication. For those who wish to try, here's a checklist.

How to "Rebloom" a Poinsettia

The first step is care. Be sure you tend your plant properly during the first holiday season with warmth, tender loving care, water, and appreciation. Never let it dry out!

By February, your poinsettia may begin to fade in color. Keep it near a sunny window, but not in direct hot sun that dries it out. By mid-April, cut stems back to approximately 6 inches above the soil. In May, start fertilizing with a balanced 20-20-20 or 20-10-20 analysis fertilizer of nitrogen, phosphorus, and potassium at 1 teaspoon per gallon of water every third watering.

By June, remove the plant and check to see if it needs repotting because of too many cramped roots in the pot. If you do repot, use commercial potting soil and a slightly larger pot. The pot may be placed outside in light shade when temperatures do not fall below 60 degrees, or keep it inside. Fertilize with 1 teaspoon of the balanced fertilizer per gallon of water every second time your water.

By mid-August, the plant should be kept inside, in direct sunlight. Cut stems back again if necessary so there are only 3 to 4 leaves per stem. Continue your watering and fertilizing cycles.

By the middle of September to the first of December, the plant should stay in direct sunlight, next to a window, until 5 P.M. From 5 P.M. to 8 A.M. the plant should be placed in complete darkness. "What do you mean?" you may ask. Here's what to do.

Put the plant in a closet or a tight box and cover it completely with a black plastic bag overnight that stops light from reaching the plant. That's what the plant needs—alternate sun and full darkness to force colorful new bracts and the tiny blooms, too. In this regard it is similar to the special requirement of sugar maple trees, warm days and cold nights to force sap flow that can be tapped and then boiled off to produce maple syrup. Plants have their own peculiar and specific needs. In the case of poinsettias they can be recycled, but you must pay attention to their very special needs to get good results.

The poinsettia is a photoperiodic plant, meaning that it sets its growth and produces flowers as the autumn nights lengthen. The plants will naturally come into full bloom during November or December, depending upon the genetic flowering of the individual cultivar.

Timing the bloom to coincide closely with the Christmas holiday can be difficult without the controlled environment of a greenhouse. Stray light of any kind, such as from outside street lights or household lamps, could delay or entirely halt the reflowering process.

Commercial growers explain that starting October 1, plants must be kept in complete darkness for 14 continuous hours each night. During October, November, and early December, the plants require six to eight hours of bright sunlight daily, with night temperatures between 60 and 70 degrees. Temperatures outside this range may delay flowering. During this time, you must continue the normal watering and fertilizer program for eight to ten weeks to achieve a colorful display of blooms for the holiday season.

Even when you succeed, the plant will not look like it did originally, or others do in stores. Although you provided all the care your plant needed following these directions, you cannot provide a professional greenhouse environment. That includes controlled lighting, temperatures, and professional growers who are tuned in to these sensitive plants. Professional gardeners can also produce a larger, fuller plant than you can based on their production facilities and experience. If you do repeat your reflowering recycling process, be aware your plant will be about 2 to 4 inches taller each year, so you should repot into a larger container as needed.

Perhaps some final thoughts or factoids about this fascinating plant may be of interest. I pass them along courtesy of friends at the Paul Eke Ranch in Encinitas, California.

The poinsettia is the Number 1–selling potted plant in America. It was originally sold as cut flowers and is now the traditional Christmas flower all around the world. Encinitas, California, is the poinsettia capital of the world.

Poinsettia Safety Research Results

The poinsettia was, as discussed, named for Joel Roberts Poinsett, the first U.S. ambassador to Mexico, and it is native to Mexico, where it was supposedly used as a medicine by the ancient Aztecs. That proves another misunderstood point. The poinsettia is not poisonous, as periodic rumors note. How did that myth begin? Supposedly in 1919 an Army officer claimed the death of his child was the result of eating poinsettia bracts. That story was later determined to be only a rumor, but the myth took root. Incorrect facts and myths have a way of lingering.

No other consumer plant has been tested for toxicity as much as the poinsettia. All research results have found no toxicity with ingestion of any part of the plant. Even so, it is still widely believed that ingestion is dangerous. During my 25+ years as a nationally syndicated columnist, I have tried to set the record straight. So have hundreds more of my garden writing colleagues and friends. Perhaps this book and chapter will serve again as a source of facts: Poinsettias are nontoxic.

The safety of poinsettias in the home was clearly demonstrated in scientific studies conducted by Ohio State University in cooperation with the Society of American Florists. The study concluded that no toxicity was evident at experimental ingestion levels which far exceeded those likely to occur in a home environment. The POISINDEX® information service, the primary information resource used by most poison control centers, states that a 50-pound child would have to ingest more than 500 poinsettia bracts to surpass experimental doses. Even at that high level, no toxicity was demonstrated.

As with all ornamental plants, however, the poinsettia is not intended for human or animal consumption, but consumers should feel at ease when decorating their home or giving these plants as gifts, according to botanists at the Paul Eke Ranch, which has been growing and breeding this holiday plant for more than 70 years.

Therefore, as can be seen from research and the fact that despite 70 million sold each year, no major poisoning problems have been in the news, it seems obvious that poinsettias can be safely displayed in the home and church as many millions have been for decades.

CHRISTMAS CACTUS AND EASTER CACTUS

Flowering cactuses as house plants are another popular and traditional holiday plant, mainly during the Christmas season when one variety comes into glorious full bloom, or Easter when the so-called Easter cactus blooms.

We've had our Christmas cactus plants for decades. The first cuttings were presented to my wife by an elder in our family. Since then, we have proudly clipped and started cuttings of these Swenson family heritage plants to give to new family members following weddings. That is now a tradition being passed along. The parent plants must be nearly 60 or more years old!

With proper care, which we'll share with you here, you, too, can bring the Christmas cactus into colorful, abundant bloom for the holiday season, right on time. This past year, by counting fallen blooms, our three basic plants graced our music room with more than 300 blossoms. A glorious holiday sight indeed.

The Christmas cactus usually flowers between Thanksgiving through Christmas. Its leaves have pointed lobes, and its botanical name is *Zygocactus truncatus*. The Easter cactus, *Schlumberger bridgessii*, has wider leaves which are rounded. It usually flowers from about Christmas to Easter. No matter which plant you have, are given, or purchase, the culture in the home is similar. Considering that ours have thrived for forty-plus years, I trust this information will be helpful to your success with them.

These plants prefer a well-drained soil, with 3 parts compost, 1 part peat moss, and 2 parts sand. You can utilize a commercial mix now available in garden centers, but be sure it is a loose soil, which cactus prefer and must have. An occasional light application of a complete fertilizer is beneficial. Consult the package of the brand you prefer to use for specific application rates. Different formulations require differing amounts, of course.

Temperature Is an Important Factor

Bud initiation normally takes place during early fall when the days become shorter and temperature is lower. To set flower buds, one of the following conditions must be met:

At 50 to 55 degrees F night temperature, flower buds will form regardless of day length. At 60 to 65 degrees F night temperature, supply thirteen uninterrupted hours of darkness, such as in an unused darkened room or closet. To put it another way, protect plants from daylight and electric lights or even streetlights from September to November. Flower buds seldom form at night temperatures above 70 degrees F. After the buds are well developed, however, they will flower at normal house temperature.

Depending on the hours of light and temperature the plant receives, the Christmas cactus should bloom during late fall or early winter. With my wife's careful timing system, which she has developed over decades, we count on glorious blooming displays every Christmas and have never been disappointed. The system works. It will for you.

Water Timing Is Important

Water regularly once a week through the end of June. Water again once at the end of July, using only the normal weekly amount of water. For a 9-inch-diameter pot that is 11 inches deep, we give it 1½ cups of water each watering. The smaller 6-inch-diameter pots get only 1¼ cup of water each watering. Water once again at the end of August, just a week's amount of water. Then, water again 3 weeks later, then 2 weeks later, then 1 week later, and you are back to weekly waterings. The weeks of no watering replicate the drought time in the desert. Sheila's system has worked well, inducing profuse blooms for the Christmas season each year.

The flower buds of the Easter cactus will develop more slowly and bloom about midwinter. After flowering, the plant will produce new growth. It is on this growth that flower buds appear for the following year. During this period of active growth, more fertilizer and water should be provided. About mid-August, reducing watering hardens off the new growth.

Withholding water slows down the plant's metabolism so that carbohydrates are stored within the plant rather than being used for new growth. From this time until flowering, the plant should be watered sparingly, only as needed to sustain it. Some experimenting may be needed for your particular plant, but the basic system works well.

These cactus plants are easily propagated from leaf cuttings at any time of the year. Simply cut off pieces about 4 to 6 inches long and place them in a glass of water. As tiny roots begin to form, pot them in a sandy soil growing mix and water enough to keep them growing. We've made new friends with these delightful, traditional plants that brighten homes every Christmas season. You can enjoy them and add them to your family holiday traditions too.

Chapter Nine

❧

Biblical Gardens to
Visit in the United States

When I first began researching Biblical gardens in 1980, only a handful of gardens seemed to exist, based on replies I received to my letters and phone calls nationwide. Happily, the idea of planting Biblical gardens has gained ground, judging from the many marvelous Biblical gardens now growing all around America. I would like to think that my talk at the Cathedral of Saint John the Divine in New York, some twenty-plus years ago, helped plant the seeds for Biblical gardens. During my talk at the Cathedral, I focused on the value of growing together with Biblical gardens, with children, family, friends, and neighbors. In a way, I offered that speech as a challenge to all who heard it and their families to dig in and plant Biblical gardens wherever they lived. My talk was well covered by the media, so perhaps millions of people read it as the wire service and syndicated stories were published in newspapers throughout the United States.

My challenge stands today and is the prologue of this book which is based on that original talk. I hope many more people will respond to the challenge. In the two decades since then, I've given many other talks and developed colorful slide programs about Biblical gardens. My goal remains the same—to encourage people of all faiths and denominations to grow together with Biblical gardens.

Thanks to the Internet, I've been able to visit with wonderful people who have been tending marvelous Biblical gardens all across America and in countries around the world too. Some are amazingly extensive, with more than 100 different plants growing in gracefully designed plantscapes. Others are more modest, but equally well researched based on Scriptural passages. Illustrations of some gardens are courtesy of those who shared the beauty of their Biblical flowers for all to enjoy. I trust that these will be an inspiration leading to creation of dozens, perhaps even hundreds more Biblical gardens of all sizes, shapes, and types in the years ahead. You can also visit many gardens at their websites, which I've included in Chapters 11 and 12. Some have extensive information and plant lists. Others have colorful pictures. Some have useful plant information that can be helpful plus links to other gardens via the Internet across America and around the world.

With appreciation to all who have planned, tended, and faithfully cared for these Biblical gardens and shared their knowledge, here's a brief tour of them. I've also included personal comments from the founders, today's dedicated tenders, curators, and directors.

FAIR HAVEN BIBLICAL GARDEN, FIRST CONGREGATIONAL CHURCH, FAIR HAVEN, VERMONT

2 North Park Place, Fair Haven, VT 05743
Tel: 802-265-8605
www.sover.net/~hkfamily

In 1981, Rev. Marsh Hudson-Knapp visited Israel for several weeks. Upon his return he developed a lively interest in the plants of the Bible. From that inspiration the Biblical garden at the First Congregational Church sprouted in his mind and at his church and has given pleasure and meaning to many. Working with his wife Cindy and other dedicated church members, he has nurtured the gardens for more than 20 years and now they contain a wide range of flowers, herbs, trees, and other representative Biblical plants. During this time, Marsh methodically searched for plant sources to make this garden truly representative of flowers and plants mentioned in the Scriptures as well as those that are native to the Holy Land. By the turn of the millennium, there were 73 different plants from flowers and herbs to vegetables, fruits, and trees as well as food crop plants in that marvelous garden.

In the process, Rev. Hudson-Knapp has written about the garden and the plants with their Scriptural roots. It seems appropriate to quote him directly because of the meaning and significance of his research, knowledge, and wisdom.

"All through the Bible, trees and flowers, fruits and vegetables play prominent parts," Rev. Hudson-Knapp says. "Every year our Jewish brothers and sisters celebrate their deliverance by observing the Feast of Passover. One part of the seder meal involves eating bitter herbs, a reminder of how bitter life was for our ancestors when they were slaves."

Dandelions have been identified as one of the bitter herbs and have a certain blooming beauty that qualifies them as a flower to be considered for inclusion in a Biblical flower garden.

Recalling the story of Gideon as told in Judges, the globe thistle is another flower featured in the Fair Haven garden. So is the star of Bethlehem which reflects the "Dove's Dung" described in II Kings 6:24–25. "And there was a great famine in Samaria: and behold, they besieged it, until an ass's head was sold for fourscore pieces of silver, and the fourth part of dove's dung for five pieces of silver."

"It appears that everyone in the Bible had their times of trial, just as we do," Marsh notes. "One of the sufferings Job experienced was the loss of appetite. In that time, hollyhock and mallow were used to flavor food. Because these flowers were used in Biblical times, they also appear in the garden.

"Plants not only played a part in the history of our ancestors, they furnished images for poetry. For example, the Song of Solomon tells a love story that models the kind of relationship God wants to have with God's people. The writer compares his beloved to the spices of a fine garden, including the precious saffron which is made from the stigmas of the beautiful saffron crocus," Marsh points out.

The Fair Haven garden blooms in season with crocus, hyacinths, tulips, daffodils, narcissus, and iris too. "The writer of Ecclesiastes also speaks of the beauty of wisdom, comparing it with the rose of Jericho. Our rugosa rose is a modern version of that simple but beautiful flower," Rev. Hudson-Knapp explains.

Jesus drew his followers' attention to the plants that bloomed abundantly around him as signs of God's abiding care. Focusing on one of the teachings, "consider the lilies of the field," Rev. Hudson-Knapp has planted an array of flowers that could have been such plants. Shasta daisies, crown anemones, ranunculus or windflowers, chrysanthemums, delphinium, hibiscus, and lupines all bloom gloriously in season.

The Fair Haven website offers much more wisdom from the Scriptures, linked with the flowers and all other plants mentioned in the Bible. It also contains links to other Biblical gardens and gardeners in America and around the world too. One of the best features is its chart that includes pictures of Biblical plants, their common and Latin names, their significance and meaning, and equally important, the appropriate Scriptural passages. The Hudson-Knapps and their congregation welcome visitors to the garden and their church in Fair Haven, Vermont. Their goal, like that of many other Biblical gardeners, is to encourage many more people of all faiths and denominations to plant and nurture Biblical gardens all across the United States.

TEMPLE BETH SHALOM BIBLICAL GARDEN— SUN CITY, ARIZONA

Jewish Community Center of Sun City, 12202 101st Avenue,
    Sun City, AZ 85351
Tel: 602-977-3240 and 972-4593
www.goodnet.com/~tbsa/body_inde.html

A Biblical garden has been thriving since 1988 at Temple Beth Shalom and the Jewish Community Center of Sun City, Arizona. Back in 1986 at a Temple board meeting the late Rabbi Bernard Kligfeld remarked: "Wouldn't it be nice to have a Biblical garden?" After nearly two years of research that garden came into being, designed, planted, and maintained by Colonel Hy Mandell, USA Rtd. Each year since the original planting, on Tu B'Shevat, a new planting is made with appropriate ceremonies, Hy Mandell explains. Each congregant is given the opportunity to add a spadeful of earth so each can say, "I planted."

"To date, forty-plus trees, bushes and other plants are growing on a three-quarter-acre garden. A sample includes palm, acacia, black and white figs, citron, almond, olive, pomegranates, carob, walnut, allepo pine, and cypress. Spices include sage, aloe, rosemary, and magnificent frankincense. Centered in the garden is an eight-foot-wide bubbling fountain and grape arbor with a small rose bed closest to the Temple," he points out.

"The guiding light in our planting is that what we put into the ground must be as identified in the Torah, commonly referred to as the Old Testament. It must be able to acclimate to the desert climate with negligible water requirements," Colonel Mandell explains. "We're constantly escorting church and other groups through the garden.

"Mountain States Wholesale Nursery has helped find several plants. Ron Gass is the Nursery Manager, P. O. Box 2500, Litchfield Park, AZ 85340. Additionally our Sun City Nursery, 9715 W. Peoria Ave., Peoria, Arizona, has helped search out many plants. My major problem now is space, finding plants to fit that require minimal maintenance."

PARADISE VALLEY UNITED METHODIST CHURCH BIBLICAL GARDEN, ARIZONA

4455 E. Lincoln Drive, Paradise Valley, AZ 85253
Tel: 602-840-8360
www.pvumc.org/about/biblicalgarden.html

The idea for a Biblical garden at Paradise Valley United Methodist Church grew from the similarity of Palestine's climate to that of Arizona, according to Bob Morgan, the tender of that delightful spot.

"The garden is located in a wash behind the original chapel and includes flowers as well as other Biblical plants. The anemone or 'lily of the field' produces a mass of red blooms in the garden around Easter each year here in Arizona, just as they cover the plains near the Sea of Galilee," Morgan points out. "Other flowers include the Bansia rose, African daisy, and oleander, a 'rose tree' found growing by the water as referred to in the Bible," he adds.

"The very word 'Palestine' means land of palms, and the date palms planted in our garden near the pool cover the hills of Palestine," gardener Morgan points out. Other key plants include olive, papyrus, myrtle, crown of thorns, olive, palm, pine, and pomegranate.

PHIPPS CONSERVATORY BIBLICAL GARDEN, PITTSBURGH, PENNSYLVANIA

613 Oxford Boulevard, Pittsburgh, PA 15243
Tel: 412-622-6916
www.phipps.conservatory.org

There are two gardens in Pittsburgh. One is the Plants of the Bible at Phipps Conservatory in Schenley Park. Its brochure notes that there are hundreds of references to trees, plants, and flowers of the countryside, most in Genesis and Isaiah and the five poetic books from Job to Song of Solomon, because plants formed an essential part of the life of the ancient Jews. That attractive Phipps garden includes crocus, lilies, tulips, and flax among the flowers. It also features figs, olives, grapes, pomegranates, and a variety of herbs mentioned in the Scriptures.

RODEF SHALOM BIBLICAL BOTANICAL GARDEN, PITTSBURGH, PENNSYLVANIA

4905 5th Avenue, Pittsburgh, PA 15213
Tel: 412-621-6566
www.rodefshalom.org/Garden/initial.html

One of the largest Biblical gardens in America, a third of an acre, is the Rodef Shalom Biblical Botanical Garden in Pittsburgh. This dramatic garden has more than 100 varieties of flora. Each plant is identified with its Biblical name or reference.

This garden is reminiscent of the Holy Land, including a cascading waterfall, a desert, and a bubbling stream that represents the Jordan River meandering from Lake Kineret to the Dead Sea. Director Irene Jacob explains that this extensive garden was begun in 1986 and presents the most complete array of Biblical plants in North America. Because a high percentage of the plants are tropical or semitropical, they are brought indoors each fall and carefully maintained through the winters, Irene Jacob explains.

This delightful garden is located in the Oakland section of the city, adjacent to Carnegie-Mellon University and the University of Pittsburgh. Knowledgeable trained docents serve the thousands of visitors who enjoy it during their garden's June to September season.

The plant world of the Bible was especially rich because of the varied topography of the land. To capture that feeling, the Rodef Shalom garden includes foods—wheat, barley, millet, and many herbs along with olives, dates, pomegranates, figs, sycamores, and others. Lotus, papyrus, rushes, and water lilies are also on display.

In addition, the Rodef Shalom Biblical Botanical Garden has sponsored an ongoing research program. Many themes have served as the focus for its annual programs, from Fragrance Through the Ages to Colors from Nature from Bible to Present. Details about the garden and program are available by calling Irene Jacob at the office: 412-621-6566.

"Plants have often been used as a quick road to understanding," according to the Rodef Shalom garden website. The website writers note that Christianity has educated people through symbols, and plants have played an important role. "A plant in the hand of Mary, Jesus, a saint, or a martyr offered a clue to the identity of the bearer in a painting. That plant depicted in the artwork then served as a daily reminder of a religious idea when seen growing in its native surrounding. Great painters also used plants, especially lilies, adding their floral and symbolic meaning to great works of art."

MAGNOLIA PLANTATION BIBLICAL GARDEN, CHARLESTON, SOUTH CAROLINA

3550 Ashley River Road, Charleston, SC 29414
Tel: 800-367-3517
www.magnoliaplantation.com/gardens/index.html

One of the most dramatic Biblical gardens in America has been created by Drayton Hastie, owner of Magnolia Plantation and Gardens at 3550 Ashley River Road in Charleston, South Carolina. That outstanding Plantation dates to 1670 and has been meticulously maintained with traditional Southern gardens by eleven family generations. It also includes one of America's most dramatic Biblical gardens, which was created following the publication of my first book, *Your Biblical Garden,* back in 1981, according to a plaque at the site.

According to Genesis, God created plants on the third day. Although it is admittedly impossible to precisely identify all Biblical plants because of lack of botanical nomenclature during the early translations, and the various different translations through the early centuries, this garden tries to provide as accurate identification as possible of the plants growing there.

As visitors explore the plants and flowers of the New Testament which surround the cross, and in the beds of the cross representing the twelve disciples, the visual impact is extraordinary. So, too, is the visual picture as one scans the gardens of the Old Testament area surrounding the Star of David, which commemorate the twelve tribes of Israel. It should be remembered when viewing these gardens in person or via a website trip, that the purpose was to create an educational garden, not an Eden, the designer points out.

"We at Magnolia Plantation and Gardens are delighted that you are writing another book on Biblical gardens," says Taylor Drayton Nelson, grandson of the present owner. "Since its inception, our Biblical garden here has been a constant source of interest and enjoyment for visitors. We regularly receive calls from people wanting information about our garden so they can develop one at their church or in their community. We are always pleased to be of help and can provide sources for Biblical plants, too," he says.

STRYBING ARBORETUM BIBLICAL GARDEN, SAN FRANCISCO, CALIFORNIA

9th Avenue at Lincoln Way, San Francisco, CA 94122
Tel: 415-661-1316
www.strybing.org

There is an exceptional Biblical plant collection in San Francisco at Strybing Arboretum. This carefully documented collection includes plants mentioned in the Bible or thought to be growing at the eastern end of the Mediterranean Sea during Biblical times. Plants then were a vital part of everyday life for food, medicine, fragrance, seasoning, and clothing. That importance to people is proved in the frequent reference to plants in the passages of the Bible. Today, many of these same plants, or close relatives, thrive in California, where the climate closely matches the summer-dry climate of the eastern Mediterranean.

The Strybing Biblical Garden features fragrant and flowering plants and has been attractively landscaped to suggest the potential use of these in home landscapes. That's a plus for all who want to better visualize how a Biblical garden should be designed for their house of worship or especially their own home landscapes. In the plants' identification, the classic reference, *Moldenke's Plants of the Bible,* was used. Many shrubs and herbs are included and worth seeing. Featured flowers include rockrose, violet, lavender, sage, anemone, crocus, autumn crocus, cyclamen, hyacinth, iris, lily, narcissus, star of Bethlehem, sternbergia, and tulip.

The Sisterhood Guild Bible Walk Fund provides for ongoing maintenance of the garden. A scholarship has been established with the proceeds from the Guild's Holland bulb sales and is given to assist the Stybring Society's teaching program. This living museum of plants is open to the public every day of the week and on weekends. Its main purpose is to educate the public about the beauty and importance of the plant world. In this magnificent 70-acre location there are more than 6,000 different kinds of plants. Docents lead tours most days for this non-profit, volunteer organization. This is one of the best botanical gardens in America and you can also enjoy a virtual walk when you click on to its website, a most colorful and informative site for all gardeners. The Strybing Arboretum Society was founded to support the development of the Arboretum and to manage the education programs.

WARSAW BIBLICAL GARDENS, WARSAW, INDIANA

P.O. Box 1223, Warsaw, IN 46580
www.warsawbiblicalgarden.org

Another dramatic site to see is Warsaw Biblical Gardens, just off busy State Road 15 in Warsaw, Indiana. It is a small plot but filled with more than 100 types of plants mentioned in the Bible. A stone wall encircles the gated garden, but it is free and open to the public. An endowment helps maintain the park, which is open from April 15 to October 15, with guided tours available from May through September for a small donation.

Happily, these lovely gardens were created on the site of a former scrapyard. The scrapyard owners, the Levin family, had the idea for the Biblical gardens. Their efforts, private donations, and support of the Warsaw Community Development Corporation saw the idea to fruition. Their website has been updated and enlarged to reflect the beauty and extent of this special spot.

In this lovely garden you'll find a gathering area, a sitting area, forest, meadow, some food- and crop-growing spots, a desert, an orchard, and a winding brook for water plants. Here in microcosm is a slice of Holy Land habitat in which Biblical plants grow, tracing their roots to the Scriptures. A unique feature is the organization of this garden in ecosystems of related plants like those in Biblical areas. Families will find the adjacent Center Lake beach a delight for children especially after touring.

In addition to providing a place of beauty, the Warsaw Biblical Gardens serve as a public research center for the study of plants of the Bible. It is educational, inspirational, and accessible to persons of all ages and all faiths as a community garden. As such, it stands as positive evidence of what a community can do to provide an uplifting educational opportunity to help people of diverse backgrounds grow better together.

TEMPLE SINAI BIBLICAL GARDEN, NEWPORT NEWS, VIRGINIA

11620 Warwick Road, Newport News, VA 23601
Tel: 757-596-8352
www.ujcvp.org/temple_sinai/bulletin04-00

Another delightful garden thrives on Virginia's tidewater peninsula at Temple Sinai. The Edward E. Kahn Memorial

Biblical Garden is replete with dozens of glorious Biblical flowers and plants. April is an ideal time to visit this garden in Newport News. Small green buds are swelling on tree branches as flowers from bulbs grace the ground.

As its website so aptly states, "It is not by coincidence that this is our Season of Redemption and Renewal. Surely it is by design that Pesach, and our Christian neighbors' observance of Easter, for that matter, occur at this time of year. This is the moment when the world around us is awakening from its winter rest."

As Curator Lee Kahn Goldfarb writes so well, "From the Garden of Eden in Genesis to the tree of life in Revelation, almost every book of the Bible contains references to flowering plants. The idea of this Biblical Garden emerged as something that might be of interest to members and friends of Temple Sinai who attend services or come to visit. It was also seen as a valuable and instructive tool for children."

From crocus and narcissus to the pale blue flowers of flax, to the anemone, iris, tulip identified with Scriptures to the blooming almond trees, this garden encompasses a wide range of plants. No doubt this attractive garden is valued as a teaching tool for children. Beyond those plants specifically identified in the Scriptures, this garden also includes others that have grown naturally for millennia in the Holy Land as they do today. You'll find buttercup, chrysanthemum, Shasta daisy, and even dandelion. That flower, oft despised by green lawn growers, has its own distinctive golden blooms and is identified as one of the bitter herbs in Scriptures.

In addition, hibiscus, larkspur, lavender, water lily, lupine, marigold, Phoenician rose, globe thistle, and violet are included. Some botanical purists may not agree with some plants on the list, but no doubt extensive research has been done.

Fact is, I must agree that those on display do indeed trace their historic roots to the Middle East and Holy Land. Considering that the Holy Land was a crossroads for caravans of ancient days, seeds and plants from many countries were passed along, traded, and planted. They took root and grow there this millennium, too. The Temple Sinai website—http://www.ujcvp.org/temple_sinai/bulletin04-00.html—to its Biblical Garden contains the full list of plants. It is worth accessing for historical as well as your own garden-planning purposes. The garden has been planted with almost 100 species of plant life that existed in the Holy Land 2000 years ago and more. The plants are identified with plaques that list the Hebrew name, common name, botanical name, and the verse in the Bible that refers to the plant, to give inspiration and pleasure to the people who tour the garden each year.

The curator of the Edward E. Kahn Biblical Garden has several key points worth repeating and remembering. Its site notes:

> Current botanical nomenclature is of comparatively recent origin and even today, occasionally, authorities disagree on the proper classification and name for certain plants. Similarly, various translations of the Scriptures use different names for plants mentioned in the Bible. In certain instances where plants of the exact species were not available, we have used available species or cultivars of the same genus. Plants are listed alphabetically by common name, followed by the botanical name and finally by scriptural reference.

I recommend that you access that Temple Sinai list as you plan your own Biblical garden growing adventure. One fact we must all honor is that the blooming beauty of flowers is one of God's gifts. If we err on some specific species or variety, that is less important than the joy we have and give by growing blooming beauty to share with others.

THE BIBLICAL GARDEN AT ST. JOHN'S EPISCOPAL CHURCH, NORMAN, OKLAHOMA

P.O. Box 2088, Norman, OK 73070
Tel: 405-321-3020
www.episcopalnorman.org/outreach.html#garden

In Oklahoma, another beautiful garden graces the church grounds at St. John's Episcopal Church. The Biblical Garden Guild was created to plan, establish, and maintain a collection of plants abundant at the time of the Old Testament. This ministry's work can be toured on the south side of the church. That lovely garden has generated many requests for information from organizations wanting to establish their own Biblical gardens, according to Director Betty Burns. She invites all church members to volunteer for a variety of tasks, from planting and weeding to ordering plants. She also kindly answers questions. More details are on the website.

HIGHLANDS PRESBYTERIAN CHURCH BIBLICAL GARDEN, GAINESVILLE, FLORIDA

1001 NE 16th Avenue, Gainesville, FL 32601
Tel: 352-376-2440
www.gnv.fdt.net~hpc/garden

Visitors can sit by a bubbling fountain and look upon many of the plants and trees mentioned throughout the Bible at the Highlands Presbyterian Church Biblical Garden in Gainesville, Florida. Its website even offers a musical interlude, "In the Garden," as you contemplate God's majesty in the quietness and seclusion of God's creation. The website has the words to that old favorite so you can sing along if you wish as you contemplate your own Biblical garden.

ST. JAMES LUTHERAN CHURCH BIBLICAL GARDEN, CORAL GABLES, FLORIDA

110 Phoenetia Avenue, Coral Gables, FL
Tel: 305-443-0014
www.st-james-church.org

Another respite from today's overly busy world is the Biblical garden at St. James Lutheran Church in Coral Gables, Florida. During the past few years, a large number of Biblical plants and shrubs have been successfully established in the Garden of Our Lord at this church. The garden is surrounded by a stone wall on which are hung large bronze plaques dedicated to nationally recognized religious, civic, and military leaders. Visitors from all parts of the United States and foreign countries visit the garden every year. It has been developed as a community and nationwide project.

According to its informative website, "The Song of Solomon lauds the costly spikenard and aloes, to be found along the paths of this garden as are hyssop, mint and cinnamon." The small bushy myrtle, used by Hebrews for perfume and spices, the lily and the bullrushes such as were used for the ark in which the infant Moses was found lying along the Nile, as well as willows are included in the garden.

At the head of the children's pool, where a lotus blooms, there is an inspiring statue of Christ, hewn in Italy from an 8-ton block of white Carrara marble. Stone benches and a vine-covered arbor complete this setting. Diagonally across from the figure of Christ, His arm raised in a gesture of benediction, is a 7-foot-high wrought iron cross, part of the garden chapel where people visit for a quiet moment of meditation. In this marvelous garden are flowers grown from seeds sent directly from the Kidron Valley in the Holy Land.

Those who have visited have come away touched by this special Biblical garden. It gives a clear picture of the everyday lives of the men and women who made Biblical history and helps us realize the important role plants have played in the Scriptures and people's lives.

THE BIBLICAL GARDEN AT TEMPLE BETH-EL, PROVIDENCE, RHODE ISLAND

70 Orchard Avenue, Providence, RI 02906
Tel: 401-331-6070
www.temple-beth-el.org

The central theme of this garden is the stone tablets of the Ten Commandments. They are framed in branches of a weeping mulberry tree that, with the help of careful pruning, droops in the shape of the tablets. Other highlights include a shaped pyracantha or firethorn in the form of the seven branched candelabra known as the Menorah, according to founder and first president of the Eden Garden Club of Temple Beth-El, Mrs. David C. Adelman.

"Because of our climate, or because of the exposed location, or because the plant I wanted for biblical authenticity happened to be an annual, or I could not obtain the exact materials I sought to achieve biblical continuity of special interest, I therefore used individual plants with symbolic backgrounds," Mrs. Adelman notes in her comments for the garden's information folder.

That makes sense to many who feel frustrated trying to obtain Biblical plants. As other gardeners have discovered, it is better to select closely related plants that will grow, thrive, and perform well, rather than risk poor growth and performance.

I agree with those who believe it is best to have at least some representative plant rather than no plant. Too often, people planning Biblical gardens had few reliable sources before my first book. This book should be a bigger help today. I have carefully searched for and included reliable plant sources. In addition, I've provided a list for you to get dozens of free mail-order plant catalogs as well as Internet sites to find plants and cultivation advice too. Use my favorite websites to Biblical gardens. The gardeners there are happy to share their knowledge and plant sources.

Mrs. Adelman wisely adds her thoughts about using attractive, alternate plants that provide their own important symbolism and beauty.

"The ground cover holds myrtle, *Vinca minor*. This is not the kind that grows in the Holy Land, which is a tree with snow white blossoms and fragrant foliage. The Hebrew name for that one is 'hadas.' Queen Esther was named for it, because of her sweetness and purity. She was named 'Hadassah.' So, our *Vinca minor* with its blue blossoms and trailing vines is the symbol of the variety grown in Israel."

In the process of creating this garden, other alternate plants were used. Instead of cedar of Lebanon, almost impossible to find today, they used small Andora junipers as a reminder of the actual species of juniper used in building temples.

To represent the "Burning Bush" from which the angel of the Lord spoke to Moses, a Winged Euonymus Alatus was selected. "This has beautiful red transparent berries in the fall, and when sun shines through them, they seem to resemble flame," Mrs. Adelman wrote in her historic notes for the Beth-El Biblical garden. The documents were discovered in the archives of the Eden Garden Club.

"I should like to say that in the creation of this garden, I humbly hope to make of the Bible a living thing, to urge contemplation of God's wonders, and His power and glory as something to be brought out beyond the doors of the Synagogue and Church—not pedantry, but Fruition!"

Presently Martha Finger has been overseeing the garden with other active members of the Eden Garden Club including president Frances Sadler, Janet Freedman, and Betty Adler.

PRINCE OF PEACE LUTHERAN CHURCH GARDEN, AUGUSTA, MAINE

209 Eastern Avenue, Augusta, ME 04330
Tel: 207-621-1768
www.poplink.org

Another well-focused Biblical garden began several years ago at the Prince of Peace Lutheran Church in Augusta, Maine. Parishioners had been discussing a project that would give church members a microcosm of what they might see in a typical Holy Land landscape. It could also provide a visual and aromatic perspective of the flowers, herbs, and other plants mentioned in the Bible.

Joseph Scott, a member of the church, took on the research and design. Using his talents and experience as a staff member at the New York Botanical Gardens and former Maine State Horticulturist, Scott was the perfect person to focus on creation of a Biblical garden. Over the years he amassed information on the climate, horticulture, and geography of the Holy Land. Concerned with accuracy, he also corresponded with Paul Steinfeld, who is affiliated with the extensive and famous Neot Kedumim Biblical Preserve in Israel.

In 1997, Scott approached Rev. David Rinas, pastor of the church, about the Biblical garden idea. The entire congregation gave hearty approval, so Scott broke ground. He prepared and enriched soil of a 20 by 24 foot plot on the front lawn. The area included rocks and a ledge similar to the hilly topography of Palestine. Today, that humble beginning has expanded. The garden includes a variety of 40 different plants native to the Holy Land. Because some species were not available, Scott had to seek alternates.

"When a substitute was made, the same genus and family was used and looks very much like the one found in the Holy Land," Scott notes. "This garden is designed to look as if you cut a chunk out of a field in the Holy Land and transplanted it to Maine."

More thoughts from Joseph Scott are included in Chapter 7, focused on designing and creating a Biblical garden with useful suggestions and advice from his experience as a veteran horticulturist.

THERAPEUTIC BIBLICAL GARDEN AT SCHERVIER NURSING CARE CENTER, RIVERDALE, NEW YORK

2975 Independence Avenue, Riverdale, NY 10463
718-548-1700
No website yet.

Charles A. Sourby has a therapeutic and healing focus with his Biblical garden. Although smaller than most, this has led to amazing and worthwhile projects for hospital and nursing home patients at health care facilities where he has worked.

"My interest in plants of the Bible has been developing for the past eight years through my work as a recreation therapist at Calvary Hospital in the Bronx, New York. I had a small Biblical garden there with 30 to 50 different species of Biblical era plants, including the Madonna lily. I used these extensively in my therapy program there. It was titled "People, Plants, & Nature" and was geared to the patient diagnosed with advanced cancer. I conducted sessions where the patients could identify, hold, and discuss various plants; refer to Biblical passages; and explore the symbolism and meaning of the same.

"The garden proved to be very symbolic of life's cycles and I believe the seasons reflected the hope of eternal life. I used and continue to use Biblical plants in my work today as a recreation director at a nursing facility in Riverdale, New York, Schervier Nursing Care Center. Recently I have worked with a landscape architect in developing a therapeutic garden in which Biblical plants will be a key ingredient. My Biblical gardens are not tidy as are many other more formal plantings. I believe the Biblical garden should reflect as natural a setting as possible, even in the confines of brick and mortar planting boxes," Sourby says.

"We are adding a therapeutic horticulture component that will include the residents using a new therapeutic sensory garden that addresses their cognitive and physical needs," he explains. "Some other ways to help people include scent gardens that are especially uplifting for the visually impaired. Many Biblical plants are richly scented. Window boxes inside a resident's window in a nursing home allow them to have their own garden even if they can't go outside. They still can pick favorite Biblical plants and provide a winter home for potted plants that will not winter outdoors in a Bible garden," Sourby points out.

Sourby's experiences have led him to focus on spirituality in a gardening elective at the New York Botanical Garden. In that course, he explores the healing of the soul that gardening and gardens offer. "We are invited to discover the inner garden of our soul. Connecting the soil, seed, and gardens leads us to the presence of God and the mysterious rhythm of healing and growth," Sourby believes. "This conceptual paradigm can be integrated into recreation activity programs in the long-term-care setting."

ALTA SIERRA BIBLICAL GARDENS, GRASS VALLEY, CALIFORNIA

16343 Auburn Road, Grass Valley, CA 95949
Tel: 530-272-1363
No website yet.

On the slopes of the Sierra Nevada foothills is nestled among aging oaks and pine trees an inspiring beauty, one of California's more unusual scenic attractions.

Alta Sierra Biblical Gardens is located on an old wagon trail near Grass Valley that was used during the California Gold Rush. It has been regraded and the trail leads to the Biblical Gardens. These gardens were opened in May 1976, the result of the inspiration of John and Verna Sommer, a Christian couple. John began creating stone statuary and wooden hand carvings to depict scenes in the dramatic story of Christ.

In 1996 another couple, Paula and Maskey Heath, were drawn to the gardens by their love of the outdoors and gardening. They admitted feeling an inner peace and tranquility in the midst of such beauty. They met the Sommers and became the owners of Alta Sierra Biblical Gardens, which they plan to operate during their time on earth as a nondenominational garden and then it will be given to the community. Today, cars full of people, busloads of school children and church groups, as well as garden clubs, artists, and individuals seeking to enrich their lives spiritually visit.

BIBLICAL GARDEN, CHURCH OF THE HOLY SPIRIT, ORLEANS, MASSACHUSETTS

Biblical Gardener—Page McMahan, 9 Morgan's Way,
   Orleans, MA 02653
Tel: 508-255-0831
www.diomass.org

Volunteers have been dedicating their efforts to this marvelous small but attractive garden for years. There are more than forty different plants from aloe and anemone to thistle and even wheat. Among the flowers of note in the list are anemone, saffron crocus, day lily, flax, iris, Madonna lily, mustard, narcissus, nasturtium, flowering quince, Provence rose and climbing rose, the star-of-Bethlehem, and globe thistle. Located on Cape Cod, this garden bids welcome to the multitudes who visit that scenic part of America's seashore.

There had been a Biblical garden, mostly of herbs, at this church for years. Recently, as part of a major parish building project, the dedicated Biblical gardeners, led by Page McMahan, expanded the garden. Although they do not have large specimen plants, they have included a wide range of Biblical plants to present a truly representative collection.

Among flowers are anemone, saffron crocus, day lily, flax, hyssop, iris, Madonna lily, mustard, narcissus, roses, both rose de

Provence and a climbing type, sage, and star of Bethlehem. A dedicated and sharing gardener, Page McMahan can be reached at 508-255-0831.

ST. GREGORY'S EPISCOPAL CHURCH BIBLICAL GARDEN, LONG BEACH, CALIFORNIA

6201 E. Willow, Long Beach, CA 90815
Tel: 562-430-1311
www.stgregoryschurch.com

This Biblical garden may be one of the more authentic and complete among those in the United States. According to co-founder Betty Clement, it has 86 plants of the Bible growing well and providing dramatic, meaningful displays to parish members and visitors alike. As a retired school principal, she has methodically researched Biblical plants for authenticity. The garden is artistically designed with a piled rock fountain in one corner, crushed rock paths stated with four impressive trees, a braided ficus, a date palm, olive, and pomegranate. Narcissis, poppies, and tulips are among the dramatic flowers. A rare Rosa phoenicia, with its simple white blooms, was obtained in England by Mona and George Nelson. Other Biblical plants grace an inner patio at the church, many in large terra-cotta pots. Markers identify them and a map is updated for visitors.

Narcissi and star-of-Bethlehem multiply themselves, but the signature red anemones and red tulips are replaced each year, Betty Clement reports. The garden is open Monday through Friday, 8:30 A.M. to 3:30 P.M., and guests are welcome to view the garden before or after Sunday services.. Information is available from Betty at 562-421-4918.

Betty Clement guides tours of the garden and does 35- to 40-minute slide presentations for church, professional, and garden groups. Her joking motto is "Have slides, will travel." Funds to support the garden come from these talks, which she offers free but most groups offer honoraria of between $25 to $200 to help support the garden. Another funding source is the annual Holly Faire booth run by the Bible Garden Guild of the church.

BIBLICAL GARDEN, OJAI PRESBYTERIAN CHURCH, OJAI, CALIFORNIA

304 North Foothill Road, Ojai, CA 93023
Tel: 805-646-1437
www.ojaipc.org

In colorful Ojai, California, a Biblical garden attracts visitors from around the state. This garden sprung from a Biblical flower show arranged by Mary Hunt in 1954. Although an Episcopalian, when the Presbyterian minister of the church suggested a Biblical garden, she dug right in. Today when visitors enter the garden, they receive a map of the grounds with names of all the plants and their marked locations. At the base of each specimen is a sign giving the plant name in Latin and the Bible references. Four Bibles were used: the King James Version, the Jerusalem Bible, the Revised Standard Version, and the New English Bible.

Trained guides give tours of the quarter-acre garden. A visit to this Biblical garden also helps guests understand Scriptural passages as they relate to actual identification of the plants. According to plant guides, the "lily of the valley" is really the blue Roman hyacinth, the "lilies of the field" are red anemones, and the rose of Isaiah is a narcissus. Tours of the garden can be arranged by calling the number, above. It is another worthwhile stop while visiting California.

THE BIBLICAL GARDEN, CATHEDRAL OF ST. JOHN THE DIVINE, NEW YORK CITY

1047 Amsterdam Avenue, New York, NY 10025
Tel: 212-678-6866
www.stjohndivine.org

This Biblical garden is one of the most impressive as it grows at the side of one of the largest cathedrals in America. Located on the 13-acre complex in which the Cathedral stands on Morningside Heights in New York City, it contains numerous species that existed in the Holy Land. Thousands of visitors marvel at its beauty and learn about the plants each year.

At the time of Christ, and even long before, most gardens were primarily functional, providing food and fruit for sustenance of the people. Gardens also were grown for their beauty and served as sanctuaries where people could find refuge, tranquility, and a place for meditation and prayer. This garden has provided a classic representation of Biblical plants since it was begun. The idea for this garden was an inspiration of the late Sarah Larkin Loening. Searching through the Bible and assorted texts about Biblical plants, Mrs. Loening realized that such a garden would provide inspiration to the thousands of pilgrims who visited the Cathedral each year and also would serve as a place for meditation and a living study garden for children. Through her leadership and with the assistance of C. Powers Taylor of Rosedale Nurseries of Hawthorne, New York, this garden was lovingly planned and planted. It was begun in 1973 and has provided inspiration to many thousands.

Nestled against the south side of the towering Cathedral walls, the quarter-acre plot inspires an ecumenical following. At times, youth groups representing Jewish and Catholic as well as Protestant schools and organizations have explored the flora of the Bible. Although it has had trees and herbs among its plants, the flowers have included anemones, narcissi, lilies, irises, and tulips that bloomed in profusion in season. All have been clearly identified with appropriate scriptural references.

Among the plants are common names, Biblical names, and botanical names. The Scriptural references listed have been taken from the book by Dr. Harold N. Moldenke and Alma L. Moldenke, one of the most reliable botanical texts about plants of the Bible, referenced to the Authorized Version of King James I published first in 1611.

This original garden was restored by Mrs. Alexander O. Victor in 1988. The restoration was planned and carried out by garden designers Mary Riley Smith and Lynden Miller and was based on a simple square-in-a-circle-in-a-square geometric design, marked on the ground with granite sets. More exotic Biblical plants such as pomegranate and date palm are kept potted outside during warm weather but are brought indoors for winter.

The guiding force and main support for the garden is the Cathedral Guild, which is headed by Mrs. Henry "Betty" Thompson. "We would like people from more churches to know about our work and to support us," she says. "This is everyone's cathedral."

As all gardens can become overgrown, this, too, has needed extensive care. A major restoration was launched at the beginning of this new millennium, according to Betty Thompson. Created by leading horticultural designer Keith Corlett, the restored garden is a spiritual and tranquil retreat that blends with the presence of the Cathedral. This was the first major renewal at the 1,600-square-foot space enclosed by fieldstone walls since it was restored in 1988 as a gift from Anna Glen Vietor in memory of her daughter. "Every flower, herb, tree, and shrub that grows there is related to plants that existed in the Holy Land 2,000 or more years ago," Mrs. Thompson points out. Visitors enter the redesigned garden through a 10-foot-tall wooden lychgate, a roofed portal that is common to churchyards. They are greeted, in appropriate seasons, by a dazzling profusion of flowers, shrubs, and trees, including crocus, iris, sage, thistle, lily, almond, and others tracing their heritage to the Holy Land.

A bluestone walk in the shape of a cross bisects the garden. At the center is a 12-foot-diameter ceramic motif with bubbling fountain. On the north side is a new addition, a rose archway that leads to a meditation walk. The Biblical garden is a haven, a secluded space to rest, or meditate, in the busy city of millions. Thousands of visitors take advantage of this oasis each year. Following the horror of September 11th, many people sought the peace of this garden to think, meditate, and pray. "I saw many people throughout the day sitting in silence in the garden," Rev. Jay Wegman said in a recent interview for *The New York Times.* As the cathedral's canon for liturgy and arts, Father Wegman often visits the garden himself. He observes, "This garden is a respite." The grounds of the world's largest Gothic cathedral are open daily from 7 A.M. to sunset. Major viewing months are May to October. Special group tours can be arranged.

BIBLICAL GARDEN, FIRST PREBYTERIAN CHURCH, NASHVILLE, TENNESSEE

4815 Franklin Pike, Nashville, TN 37220
Tel: 615-383-1815
www.klis.com/scove/041400.htm and www.fpcnashville.org

Grace Westlake moved to Fairfield Glade, Tennessee, in 1991, a planned retirement community on the Cumberland Plateau. One of the first people she met was Key Greene, an herb gardener and member of the Community Church, affiliated with the Presbyterian and United Church of Christ. Their interest in Biblical plants led naturally to creation of a garden. Today, the garden includes herbs in a wheel-shaped bed, a triad garden with seven beds including flowers that bloom in spring, yellow flag iris, a variety of hyacinths, and vegetables added for summer.

A cross is the focal point of the garden, forming four beds with biblical flowers from Madonna lilies to anemones with some blue flax, mallow, loosestrife, and tulips, too. A ladder garden with other flowers from narcissi to sweet flag, larkspur, and lilies bloom in that special spot.

BIBLICAL GARDEN OF CONGREGATION B'NAI SHALOM, WALNUT CREEK, CALIFORNIA

74 Eckley Lane, Walnut Creek, CA 94596
Tel: 925-934-9446
www.biblicalgardens.com and www.jfed.org

Avid veteran gardener Shirley Pinchev Sidell is called a visionary b; those who know her. She not only has helped create a marvelous Biblical garden at her temple, but has expanded her growing and sharing horizons around the country. The synagogue is located on an 8-acre hilltop in Walnut Creek, California. She conceived the Biblical garden during a trip to the Holy Land several years ago. Most of the plants she saw were familiar because "we share similar growing conditions and many of the same plants are found in my garden," she notes.

At that time, Shirley Sidell was chair of the landscape committee. She wanted to begin a Biblical garden but could not find adequate resources. As she continued research, she began assembling a database, which led into another productive direction. She founded biblicalgardens.com with a vision to help "build a Bible garden in every home, church, synagogue, or temple using the Internet," she explains. In the past two years gardening friends have donated millions of seeds, while others spent hours cleaning and sorting seeds for her free seed offer via her website.

The B'nai Shalom Synagogue Biblical Garden was scheduled to be dedicated on September 16, 2001, the evening just before the start of the Jewish High Holy Days.

In keeping with her focus on accuracy and shared information, all plants are well identified with descriptions and botanical as well as Scriptural references. More details about this garden and Shirley Sidell's website programs appear in Chapter 7 to help others develop their own plans, projects, and programs using some of her extensive on-line data. Seed kits also are available for sale. You can obtain information via www.jfed.org/bnaishalom.

SHIR AMI BIBLICAL GARDEN, BUCKS COUNTY JEWISH CONGREGATION, NEWTOWN, PENNSYLVANIA

101 Richboro Road, Newtown, PA 18940
Tel: 215-946-3859
www.uahc.org/congs/pa/shirami

Some gardens begin small and have a way of taking root and generating enthusiasm. In Pennsylvania, the Shir Ami Biblical garden has sprouted well, nursed along by Jules Hyman and Stan Averbach of Shir Ami Bucks Country Jewish Congregation, located in Newtown, Pennsylvania. What began on a small scale has grown magnificently; it is now 30 feet by 650 feet long!

This garden began in 1999 and has been making marvelous progress ever since. Their regular garden is called *Shomrei Gan*, which means "Keepers of the Garden." It is 30 by 400 feet, laid out with a winding path of pea-sized stones over heavy plastic sheeting and sides to minimize weeds. The path symbolizes that path of life that has many turns and can move in mysterious ways. The plants, shrubs, and trees are planted in the hollows of the path, and where possible, in Biblical order starting with Genesis. At the beginning are two cedars of Lebanon, which symbolize the two columns at the entrance of the Great Temple, which were made from the cedars.

The plantings continue following a path through the Bible. A fig grove, an apricot grove, and an apple grove are included, with a pine tree, a walnut tree, and an oak tree. This garden also contains wheat, barley, sorghum, millet, flax, and cotton in Earth Box planters to control growth. The children's garden is adjacent to the synagogue school and is 8 by 100 feet. Flowers, vegetables, and herbs are planted in Earth Boxes and large planters are placed in pairs parallel to the walkway.

All plants have dedicated markers and a pamphlet serves as a guide for visitors. Docents serve as guides for groups and the gardens are open to all without charge, but voluntary contributions are accepted and applied to the maintenance fund. Every plant has or will have a marker with the name of the plant, its Latin name, and its Hebrew name.

By 2001 there were fifty-six trees and shrubs and four grape-vines planted. Six planters contain four Biblical grains plus cotton and flax. A 5 by 15-foot pond is planned to accommodate bullrushes and papyrus. Flowers, vegetables and herbs are featured in the children's gardens, according to Stan Averbach. Later in 2001 they received six more trees and two shrubs plus a kidney shaped artificial pond to hold such authentic plants as bullrushes and papyrus. For winter, these sensitive plants, citron trees, and dwarf date palms will be moved in urns into the synagogue atrium, which is glass and faces south. It is an ongoing, creative, spiritual project and the gardeners graciously share their knowledge. Stan Averbach can be reached at Poppopstan@aol.com and loves to exchange Biblical garden knowledge, tips, and advice.

CONCORDIA LUTHERAN CHURCH, MANCHESTER, CONNECTICUT

40 Pitkin Street, Manchester, CT 06040
Tel: 860-649-5311
www.macc-ct.org/member_churches.htm

A marvelous, extensive garden with hundreds of flowers graces the grounds of Concordia Lutheran Church, according to enthusiastic gardener Jan McGarity. As the chairman of a dedicated board of active trustees, she describes the garden as a Memorial Garden that has some Biblical plants. There is a Pathway of Faith with engraved 24- by 24-inch stones, that have names of the donor supporters of the garden. There is an outdoor altar and services as well as baptisms have been held in this extensive garden. Jan McGarity is justly proud of their garden and shares information and ideas. She can be reached via e-mail at Bobmcgar@aol.com or by phone at 860-872-6266.

CHURCH OF THE WAYFARER BIBLICAL GARDEN, CARMEL-BY-THE-SEA, CALIFORNIA

Corner of Lincoln & 7th, Carmel-by-the-Sea, CA 93921
Tel: 831-624-3550
www.churchofthewayfarer.com

This marvelous and deeply rooted garden began about 1940 with Mrs. George Beardsley's inspiration and generosity. Many people have contributed to its care and growth, notably Mrs. Blanchard Steeves, who also edited its garden booklet, and Mrs. James Bishop, who helped in the original design. Mrs. Raymond Chrisman generously provided funds for many plants. Dedicated gardeners Carolina Bayne, Helene Brown, and Tracie Bayne are among the active master gardeners tending this impressive garden in the new millennium.

The village of Carmel and its Methodist church, The Church of the Wayfarer, have a long history in California. Carmelite friars marveled at the "Garden Land," a translation of the Hebrew *carmel*, which they saw as a mirror of Mount Carmel of the Holy Land. Thus, that area was named. Because of the relationship in loveliness and name, "Garden Land," it seems appropriate that the church Biblical garden should be named "The Master's Garden." From the church door, one can enter the garden to enjoy the flowers, herbs, trees, and other Biblical plants. A detailed booklet with Scriptural passages and information about all plants is available for all visitors to learn about each plant and its place in the Holy Land and Bible. One key phrase stands out: "For flowers are the poetry of Christ." A most fitting tribute to the faith of many who have faithfully tended this beautiful garden throughout the decades.

The United Methodist Church designates one Sunday each year as a festival of God's Creation or Earth Sabbath. The Biblical garden is one way they have chosen to enable congregations to celebrate God's gracious work in creating the earth and all living things that dwell upon it.

In a foreword to its garden booklet describing the garden with Scriptural references to the plants growing there, Charles C. Anker adds his thoughts, which we can all take to heart. "God planted a garden for the first man and women to live in and it was there that he walked and talked with Adam and Eve," he observes. "The garden then is the place reminding us always of God's incredible love for us. I hope that this garden reminds you always of the special love that God has for each one of you," Anker concludes.

OTHER BIBLICAL GARDENS GROWING, SPROUTING, BEING PLANNED

In this year of our new millennium, I'm pleased that there are other Biblical gardens growing, being planted, and planned. Good friend and exceptional Biblical gardener Rev. Marsh Hudson-Knapp advises that a garden club at Liberty Presbyterian Church, a country church about 12 miles north of downtown Columbus, Ohio, has started work on a Biblical garden. Marvin Languis has been gathering information for its designer, Tom Wood, to use. The contact is libertychurch@sprynet.com.

Another garden is being planned at the First Congregational Church in Winter Park, Florida. Rev. Ken Crossman, a retired United Methodist pastor, is working with Pastor Bryan Fulwider to plan a garden to launch in 2002, Marsh tells me. He also noted that Mary Jo Gibson is working with her son on an Eagle Scout project restoring Biblical plants in the church garden at St. James Lutheran Church near Wapwallopen in Pennsylvania and that Mary Lou Froehl of Saints Peter and Paul Church in Petersburg, Indiana, is also planning a new garden there. In addition, Marsh has posted other good news that the Parkside Lutheran Church in Buffalo, New York, is hoping to install a Biblical garden with the encouragement of one of its deacons. In Greenville, Texas, a master gardener, "Pud" Kearns, has designed a breathtaking Biblical garden right in the center of the city. You can see that dramatic design and get some ideas from it by visiting the Fair Haven website, where Rev. Hudson-Knapp has posted this garden plan in graphic color.

Another contact for youth gardens is Bob Puchra, whom I understand is developing the Biblical garden at Father Flanagan's Boys Home. The address is 318 Bucher Drive, Boys Town, NE 68010.

It is my hope that this book, and the inspired gardens you'll read about in it, will encourage many more people like yourself to dig in and begin growing Biblical gardens in the years ahead. Growing together with God is a worthwhile goal for every gardener.

Chapter Ten

❧

Magnificent Neot Kedumim— The Biblical Landscape Reserve and Other Biblical Gardens Worldwide

NEOT KEDUMIM, LOD, ISRAEL

From a city windowbox display of a few Biblical flowers to a modest 20- by 20-foot garden with forty Biblical plants to a reserve that covers 625 acres and re-creates the flora and fauna of the ancient Holy Land, I have seen Biblical gardens in all sizes and shapes. In this book you'll find growing ideas and opportunities to create your own world of blooming beauty as space, budget, and energy allow.

Neot Kedumim, the 625-acre Biblical Landscape Reserve in the Holy Land, is the most remarkable and magnificent Biblical plant garden in the entire world. It is more extensive than any other Biblical garden, has a wider range of authentic Biblical plants and landmarks, and presents a restored natural environment and plant habitat as gardens once were in Biblical times in the Holy Land. Because it is such a Biblical plant landmark in the world with such marvelous and meaningful outreach programs of classes, courses, and literature, and serves people of all faiths, it is appropriate to focus on it in this special chapter. There are Biblical lessons growing there that are eminently transplantable to other gardens, even small backyard plots all around the world.

Neot Kedumim began as a dream shared by Dr. Ephraim and

Hannah Hareuveni, two Russian Jewish emigrants to Israel who envisioned these gardens of flora and fauna. Both were trained botanists who dedicated themselves to research of the land and ancient literature of Israel.

They conceived the idea of developing a living replica to reflect the interrelated botany, history, and traditions of the land of the Bible, which would create a bond between the past, present, and future. The Biblical Landscape Reserve in Israel thrives today and exists on a far grander scale than they may have envisioned. It stands on 625 formerly barren acres in Israel's Modi'in region as a pastoral network of Biblical and Talmudic landscapes which attracts more than 100,000 visitors a year.

As Helen Frenkley, the director of Neot Kedumim has observed, it conveys messages through accounts of people interacting with a particular land. The language of the Scriptures is alive with sights, sounds, fragrance, and graphic visual perspectives of the Holy Land's natural landscapes. It is through such images, which are literally rooted in the Scriptures and landscapes, that the Bible conveys its ideas.

The word *Neot* means "pastures" or "places of beauty," as in Psalm 23, "He maketh me to lie down in green pastures." The word *Kedumim* means "ancient" and contains the Hebrew roots of this word that indicates forward movement in time and also expresses hope of future growth from past roots. That is a most apt and expressive name for this remarkable Biblical landscape.

Appropriately, this sunny 625-acre botanical park that reproduces Biblical terrain includes the seven varieties of food that Moses mentions in Deuteronomy 8:8 as the most important crops of ancient Israel: wheat, barley, figs, pomegranates, grapes, olives, and dates. All plants in the reserve are well marked with appropriate quotes from the Scriptures. Walking the sloping hills and paths, visitors see gleaming man-made lakes, old Byzantine cisterns and chapels, an ancient Roman olive oil press, and other relics of two millennia ago. Because it is a reserve, representative animals of the Scriptures reside there again. A camel, gazelles, sheep, goats, and cattle wander, and ducks paddle on the tranquil Pool of Solomon.

You can rest under an elegant, historic cedar of Lebanon tree. You can inhale the scents from the flowers, the herbs, the essence of earliest times. Visitors can walk several self-guided trails and most are handicapped accessible. These trails provide ideas for all who wish to explore the outreach potential of Biblical gardens at their churches or temples or organizations.

In one walk you find the fields brimming with flowers, which are an array of the wild and wonderful flowers of the Holy Land. These are a natural sight and grow well because they are native to that land. Providing a new, similar soil habitat to native plants is one way to help wild plants grow well. Native plants are accustomed to the land. That is true in the United States, where iris and daisies and similar plants just naturally take root and thrive. Providing conditions as close as possible to their native habitat is the best way to ensure that windflowers take root, thrive, bloom, and multiply.

Along one tour at Neot Kedumim you'll also see several varieties of iris and an array of cyclamens. Daisies and Charon tulips grace the pathway. Anemones, a classic flower of the Scriptures and Holy Land, provide their red, purple, and sometimes white delights. As part of this re-created Biblical Reserve, visitors may see a Gallein wolf, fallow deer, or even a camel near the water cistern.

Another trail guides visitors to vistas of plants in the Song of Songs. White lilies reflect the passages, "his lips are like lilies, dripping flowing myrrh," from Song of Songs 5:13.

Along the way are red buttercups, from Song of Songs 2:12, "The red buttercups appear on the earth." Farther along, people see grapevines and striking clusters of white and yellow narcissus, that reflect that Biblical image, "I am the narcissus of the valleys," from Song of Songs 2:1.

In the fall, there is an ingathering, a harvesting time for crops and preparing for the coming rainy season at Neot Kedumim. Visitors see figs and grapes being sun dried to preserve them as energy-rich food, as was done by the people there thousands of years ago. Fields are plowed and sown with seed grain in anticipation of the coming winter rains.

This, of course, marks the season of the ingathering, the end of summer. In Hebrew it is *kayitz,* which comes from the same root as two other ancient Hebrew words, *ketz* and *kotz. Ketz* means "end," because plants die or go dormant in the brutal heat and dryness of Israel's summer months. *Kotz* means "thorn," the kind of plants that can dominate Israel's fields at this drying time of year. As their posted plant descriptions say, even thorns can be beautiful as the purple thistles that we can grow as replicas in our gardens.

White squill, meadow saffron, which is a type of crocus, and other flowers, herbs, and trees are part of this and other hiking and walking trails, well described in markers about the plants. From these trails, printed flyers in various languages enable visitors to further enjoy and understand the important Scriptural, historical, and ecological significance of what they see growing.

Neot Kedumim continues to grow in many worthwhile directions today, a tribute to the Hareuvenis' vision and especially the dedication and perseverance of their son, Nogah Hareuveni. He was the actual founder of Neot Kedumim, the Biblical Landscape Reserve, in 1965. He was born in Jerusalem in 1924. From childhood he worked with his parents, the founders of the Museum of Biblical Botany at the Hebrew University. For years he traveled afield throughout Israel with them as they collected and recorded the vast flora of the land of the Bible, so important to several of the world's religions.

Neot Kedumim remains dedicated to exploring and demonstrating the ties between the Biblical tradition and Israel's nature and agriculture as expressed in prayers, holidays, and symbols. Equally significant, it has deep roots in Judeo-Christian traditions. Today it has a staff of sixty and hosts tens of thousands of visitors of Jewish and Christian faith annually.

Years of devotion to this exceptional, living, and growing Biblical landscape project led Nogah to write several distinctive books. His books include *Nature in Our Biblical Heritage,* the perfect gift for anyone interested in the Bible, nature, history, or Judeo-Christian traditions. *Tree and Shrub in Our Biblical Heritage,* and *Desert and Shepherd in Our Biblical Heritage* are equally worthwhile. All are available from American Friends of Neot Kedumim, listed on page 241.

When Biblical garden enthusiasts travel to this actual reserve, in person or via the Internet website, they can wander among groves of palm, fig, and olive trees, observing the intertwining of Jewish and Christian traditions. In these extensive gardens, olive trees stand symbolically as they have since ancient times. According to universal tradition, olive branches are a symbol of peace, and they are included in the state emblem of Israel. In Exodus 27:20, God commanded that olive oil was to be used in kindling the light of the menorah because it provided the brightest and steadiest flame.

Also prominent in the gardens is the moriah plant, which is a member of the sage family. This herb, which grows virtually wild throughout Israel today as it did in Biblical times, has special significance to the Jewish people. The menorah is first mentioned in the Bible when God instructed Moses in the preparation of the Ark of the Covenant. As described in Exodus 25:31–40, which you can read, the specifications seem almost couched in botanical terms of branches, calyxes, cups, and petals. Ephraim and Hannah Hareuveni were the first to point to a direct relationship between the menorah and the moriah or sage plant as a particular Biblical plant. The moriah may not always have seven branches, but it does have an even number growing from a central branch, and its pattern is strikingly similar to the menorah. Because sage plants do have flowers as well as herbal attributes, I have included them in this book with growing advice about them in Chapter 3.

Neot Kedumim is located in an area known in the Bible as the Judean foothills or lowlands. That area lies between the flat coastal plain that comprises the western edge of Israel and the Judean hills that rise toward Jerusalem to the southeast. The typical landscape of these hills, without care, is rather dry, barren and seemingly inhospitable, unless you have a vision and the determination to make this land bloom again as it had done aeons ago. That is one of the aspects that make Neot Kedumim so remarkable, that people with vision and historical facts to foresee the potential could and did make a garden bloom in what had become, over centuries of neglect, a desolate area. Centuries of overgrazing, battles, and neglect of the land had eroded it down to bedrock. To create gardens, thousands of tons of soil had to be trucked in to the site.

BIBLICAL GARDEN EDUCATIONAL AND OUTREACH PROGRAMS

The Holy Land is a microcosm of diversity in nature. Neot Kedumim itself is a microcosm of a microcosm. This 625-acre reserve re-creates a series of landscapes that are found in Israel and explores the intertwining between the land of the Bible and the texts of the Hebrew Bible and the Christian New Testament. Lectures and programs shed light on this important linkage of Biblical Scriptures and natural history. Neot Kedumim is an on-going educational project based on the tenet that many Jewish and Christian traditions, holidays, and symbols are rooted in the ecology, ancient agriculture, flora, and fauna of the Land of Israel, the Holy Land.

This garden network containing plants and animals mentioned in the Bible provides visitors with a living panorama of Judeo-Christian history and an explanation of its roots in the physical land. The original concept continues to grow and expand. On-site educational programs are conducted regularly and outreach program continues to grow as the gardens do.

More than a decade ago, Neot Kedumim became part of Israel's educational system. More than 60,000 pupils from kindergarten through high school and all parts of the country's religious and social spectrum have been visiting Neot Kedumim annually during the past decade. In addition, Neot Kedumim has reached classrooms via the hundreds of teachers who participated in its teacher enrichment and in-service training programs. Teachers from all parts of the country, religious and secular, young and old, regularly take part in educational programs. The programs have at times transcended religious differences as well. Four Moslem Arab teachers from a village in the Galilee participated in one of the courses. The following year they brought some of their colleagues to the program. Growing together in peace is a worthwhile theme at Neot Kedumim, in the Holy Land, and around the world and is the underlying theme in this book too.

The innovative and wide-ranging courses at Neot Kedumim provide programs about ecology, conservation, and the environment, which are all-important topics in an area where land preservation is vital. Other courses deal with parables and symbols in the Bible and with archeology. In addition to formal courses, both students and teachers often come in search of resource materials on the "green archeology" that Neot Kedumim uniquely offers. Courses are open to teachers and students from other countries in English, Hebrew, and Russian.

Other Biblical gardeners have developed outreach programs in their own way. Joseph Scott of Prince of Peace Lutheran Church in Maine has colorful, informative slide show programs that he gives periodically. From these he has met people of diverse backgrounds and different denominations and faiths. As this book goes to press, he has begun helping several other religious groups in Maine plan their own gardens, locate appropriate plants, and come up with fund-raising programs too.

Many people creating and sustaining Biblical gardens are also concerned with funding and with integrating the viewing of flowers with other aspects of Biblical and everyday life. The folks at Neot Kedumim do both by serving meals created with foods and decorated by flowers grown in Biblical gardens. The meals allow visitors to "taste and see" Biblical foods. The proceeds from meals also help to fund operations much like church dinners help support religious programs, groups, and gardens in the United States.

FOODS AND TRADITIONAL FEASTS

Church suppers have been and are a traditional fund-raising activity for thousands of churches across America. Perhaps some of the special food and feast projects at this innovative Biblical garden will provide food for thought as fund-raiser projects for churches and temples here in the United States. After all, foods that trace their history to Biblical plants can have special meaning for special events. Neot Kedumim folks realized that visitors get hungry. They have come up with tasty meals to feed the multitudes, or a few visitors at a time, as the case may be.

From Proverbs 15:17 we read, "Better a meal of vegetables where there is love than a fatted ox where there is hatred." In keeping with that passage, Neot Kedumim offers vegetarian menus. You may not wish to become that deeply involved as operating a regular food service at your church or temple, but there may be some opportunities for special events, fund raisers, or Holy Day suppers that come to mind as you read its menus.

Innovative ideas deserve a salute. Here is part of a typical menu from Neot Kedumim.

Before dinner a visitor can savor appetizers that include: garbanzo beans with olives in hyssop, cucumber strips in tehina, and roasted pita with olive oil and garlic.

Then, they may choose from the buffet among flavored cheese: natural, creamed, with olives, walnuts, dill, or onion, plus four kinds of fresh bread, reflective of the grains grown there now and historically. Also, flavored butters, natural, dill, onion or garlic, and pickled vegetables including a choice of cucumbers, onions, turnips, and mixed veggies.

Also included are salads: pickled quail eggs with herbs, yogurt, and cucumber strips with dill and mint, or tabouleh (which is cracked wheat with herbs), vegetables, and olive oil. Other choices are garbanzo beans with capers, chopped olives with parsley, garlic, lemon squares, and olive oil; spinach leaves with red lentils, garlic, olive oil, and mint; and sliced cucumbers with pickles, dill, and purple onion.

Hot dishes include squash and onion casserole with rice and lentils and thick lentil stew. Squash stuffed with rice and raisins and squash stuffed with cheese also are available. Among other supper treats are herb salad with parsley, coriander, mint, dill, and pine nuts; strips of squash, lightly cooked with hot and sour sauce; squash with onions and herbs; white beans with chopped spinach and onion rings in olive oil and lemon; lettuce salad, cabbage, and sesame in sweet and sour sauce; raw carrots with walnuts and raisins; beets with walnuts and herbs; grape leaves stuffed with rice.

Desserts include in-season fruit, or dough puffs with a choice of homemade jams and honey. Beverages are cold lemon and almond drinks and hot herbal tea. Wine is optional.

For those who wish a morning treat, a Biblical-style breakfast is offered as a buffet. It includes shepherd cheese, farmer's cheese, garden cucumbers, olives and olive oil, hyssop seasoning, fresh-baked pita bread, along with yogurt, date honey, bee honey, raisins, walnuts, and seasonal herbal teas. As noted in its literature, the menu seems appropriate to quote from Psalm 78:29, "and they ate and were well filled."

This delightful array of menu items offers lots of ideas for Biblical meals for fund-raising events at churches and temples with Biblical gardens here in North America. In fact, these creative menus invoke ideas for special Biblical meals, whether your church has a Biblical garden or not. For example, the First Congregational Church in Fair Haven, Vermont, includes one or more Biblical food items in a more traditional church supper as a special highlight. As the season allows, Biblical flowers are used for table decoration.

WEDDINGS AND SPECIAL EVENTS ARE POPULAR

Another idea has taken root at Neot Kedumim and become increasingly popular, having weddings along the Wedding Trail there. The Wedding Trail is a gentle trail through landscapes evoked in the Song of Songs 8:7. Proceeding from a pool reminding participants that "torrents of water cannot quench love, nor rivers drown it," the trail skirts young pomegranate trees, reflecting another Scriptural passage, "Your limbs are an orchard of pomegranates, with all choicest fruits."

After walking through the landscapes of the "tulip of

Charon," the "apple among the trees of the forest," and the "narcissus among the thorns," wedding guests arrive under the tall date palms beside the Pool of Solomon. The trail is illuminated at night and provides a truly romantic setting for a wedding. After the ceremony, the Neot Kedumim tour trail takes wedding guests to a reception area, a spot-lit clearing in a pine forest for eating, singing, and dancing. That, too, has its Scriptural reflection from Isaiah 55:12, "For you shall go out with joy, and be led forth with peace; the mountains and the hills shall break forth before you into singing, and all the trees of the field shall clap their hands."

Your Biblical garden can serve as a site for weddings or other special events. It might also serve as a special site for wedding photos, with flowers and plants selected that may have a special significance to the event or people involved.

Another aspect of Neot Kedumim deserves special focus, again for ideas that may be useful as we see how others have overcome obstacles, and creatively planted and cultivated their gardens. According to Shlomo Teitelbaum, Neot Kedumim Deputy Director for Development, the planners there faced some dilemmas while reconstructing the landscapes of the Song of Songs and had to come up with creative solutions.

"It was King Solomon's words in Ecclesiastes that gave us the inspiration and courage to develop a portion of the bare, rocky tract to the Song of Songs. Two hills hugging a valley of varying width would be ideal for verses from the Song: 'His aspect is like Lebanon, noble as the cedars,' and 'I went down to the walnut garden to see the budding inside the stream banks.' and 'you stand like a date palm.'

"But how could we plant cedars and walnut trees which grow in high cold habitats, next to date palms and the narcissus of the valleys which need warmth and large quantities of water?" Teitelbaum asked rhetorically. "How could we put 'mountains of spices' next to the Charon tulip, grazing areas for animals next to cultivated gardens of grapevines and pomegranates?

"Solutions to some of the problems were suggested by the words of the Song itself: 'I went down to the walnut garden.'

Walnut trees normally grow at high elevations, but as the verb 'went down' suggests, walnuts also can grow in a microclimate created by a topographical depression surrounded by hills. The cold air sinking into such a pocket provides nights of winter frost which the walnuts need.

"Red sandy loam was brought from the Charon plain. The topsoil, with the seeds of windflowers in it, was taken from an empty field just before apartment towers were built there. Along with the white broom and oaks of the Charon habitat, we planted bulbs of the Charon tulip.

"A pond was dug at the edge of the valley to catch runoff rainwater and provide a source for irrigating the walnut trees. Today, that vision has come to fruition and visitors can enjoy delicious ripe walnuts in the fall. Next to the pool, a large catchment basin was dug. Muddy soil that forms there during the rainy season provides a good habitat for 'narcissus among the thorns,'" Teitelbaum points out.

The pool, called appropriately the "Pool of Solomon," reflects the tall date palms planted beside it, as well as the cedars growing on top of the adjacent hill; the majestic trees that appear together in the lovers' words in the Song and in Psalm 92: "The righteous will flourish like the date palm, and grow tall like the cedar of Lebanon." "And quite naturally, this is the setting for wedding ceremonies at Neot Kedumim," Teitelbaum concludes.

Neot Kedumim is located on Route 443 near Modi'in and the Ben Shemen Forest, only ten minutes by car from Ben Gurion Airport and thirty to forty-five minutes by car from Jerusalem or Tel Aviv. Visiting hours are Sunday to Thursday, 8:30 A.M. to sunset, Friday and Jewish holiday eves 8:30 A.M. to 1 P.M. It is closed Saturdays and Jewish holidays. The international phone number is 972-8-977-0777 and the fax number is 972-8-977-0766. E-mail information is gen_info@neot-kedumim.org.il, and its colorful website, one of my favorite Biblical garden websites, is www.neot-kedumim.org.il. On that website we can all enjoy a virtual walk into Biblical history.

Until a Biblical village hotel is built at Neot Kedumim, and one is on the planning board, it is best to stay in Jerusalem or Tel

Aviv hotels. Travel agents can provide details about trips to the reserve and elsewhere in the Holy Land, as can Israel's national airline, El Al. As part of its long-range educational work, Neot Kedumim holds seminars and tailor-made tours for special-interest groups.

The best way to get information in North America is from Innes Kasanof, Executive Director of American Friends of Neot Kedumim, Steinfeld Road, Halcott Center, NY 12430. She also has copies of Nogah's books available for sale. The telephone is 845-254-5031, the fax is 845-254-9836, and her e-mail is afnk@ catskill.net. I trust that this book will also help encourage support for that marvelous project. As the olive branch stands for peace, may Neot Kedumim continue to grow and encourage more Biblical gardens to sprout and thrive in peace around the world in keeping also with the theme of this book, *Let's Grow Together.*

OTHER BIBLICAL GARDENS AROUND THE WORLD

In Jerusalem there is another garden, but unlike any you may have visited. This World of the Bible Gardens has full-scale archeological replicas that help interpret the Scriptures. There is a goathair tent, a real sheepfold, a stone manger, and a well from which visitors can draw water.

There is also a watchtower, a wine press, a threshing floor, and an olive press. At the stone quarry, visitors can learn about ancient building methods. The Scripture Garden is located in Ein Karem, Jerusalem, and offers classes and even has authentic Biblical meals. With all these attractions, the plus is the plants that are growing at this unusual site.

THE GARDEN OF GETHSEMENE

Another special garden in the Holy Land is worth noting. It is located at the western base of the Mount of Olives. Within the walled perimeter of the Church of All Nations, you can find a small grove of ancient olive trees. The historic Garden of Gethsemene holds a place in the hearts of all Christians as the site where Jesus prayed the night before he was taken captive, as described in Mark 14:32–50. In this churchyard, local tradition dates these ancient trees back to Biblical times, but that is unlikely. More probable is that the gnarled and twisted olive trees may be descendants of the original ones. They are undoubtedly very old. Although not a typical Biblical garden, this unique spot deserves its place in this book as a special garden landmark in the Holy Land.

For those who wish to wander around the world to visit other fine Biblical gardens, here are some that I've located during my research for this book. A few had websites at the time of publication and I've included them in that reference section in Chapter 12.

MORE BIBLICAL GARDENS AROUND THE WORLD

More Biblical gardens seem to be taking root every year, so I plan to add to the list in future updated editions. Meanwhile, here are Biblical gardens to know about that I've been told are growing around the world. In future editions of this book I hope to add even more and share details about them.

Yad-Hasmona Biblical Gardens, Israel
Biblical Garden, St. George's College, Jerusalem, Israel
Biblical Garden, Brickman's Country Gardens, Ontario, Canada
Bible Garden Memorial Trust, Palm Beach, New South Wales, Australia
Biblical Garden, St. Paul's Cathedral, Sydney, New South Wales, Australia
Biblical Garden, Rockhampton Botanic Garden, Queensland, Australia
Biblical Garden, Rutland Street Chapel, Christchurch, New Zealand

Biblical Garden, Holy Trinity Parish Church, Sheen Park, Richmond, Surrey, England

Millennium Bible Garden, Woodbridge, Suffolk, England

Biblical Garden, Royal Botanic Garden, Kew Gardens, Richmond, Surrey, England

Millennium Bible Garden, St. Mark's, Isle of Man, England

Biblical Garden, Sternberg Centre for Judaism, The Manor House, London N3 2SY

Biblical Garden, Bangor Cathedral Close, Bangor, Caernarvon, Wales

Biblical Garden, Elgin, Scotland

Biblical Garden, St. Benedict's Priory, The Mount, Cobb, Co. Cork, Ireland

Biblical Garden, Ara Virdis Project

Franciscan Center of Environmental Studies, Rome, Italy

Biblical Garden, Amsterdam Free University, Amsterdam, Holland

Chapter Eleven

Greatest Global Gardening
Sources for the Millennium

While you are trying to grow better with God in your Biblical garden, it's nice to know there's lots of help available whenever you need it. I've compiled my best gardening information sources for you in this book, from all across America and around the world. Even better, you can tap top gardening talent at dozens of colorful, informative websites on the Internet and print out pages of valuable garden ideas, tips, and advice. If you haven't surfed the web in search of great gardening ideas and advice, you're missing bushels of useful information and hundreds of free pages about every aspect and type of gardening. You also get to take a peek at beautiful flowers that you may wish to grow, and learn about their habits and best tips for growing each.

The Mailorder Gardening Association is the world's largest group of companies that specialize in providing garden products via mail order and on-line. At MGA's periodically revised website, www.mailordergardening.com, you can find a glossary of gardening words and phrases. It also has more than 130 members who offer colorful, illustrated catalogs packed with tips and ideas.

The best thing about this site is the impressive list of member companies. They're categorized by the types of products they provide—annuals, perennials, fruit trees, garden supplies, fertilizer. There are 26 companies listed under "bulbs" alone!

Click onto a company name and up pops a short description of the catalog's offerings, phone number, and address. Click on the hotlink and you're instantly connected to the catalog's homepage. This website is a great place to start when you want to see just how many garden seed, plant, product, and accessory firms are on-line. You also can find toll-free numbers of companies that offer personal advice to your own gardening questions.

The National Gardening Association, www.garden.org, publisher of the *National Gardening* magazine, has one of the most comprehensive and useful sites on the web. Hot buttons let you read articles from the magazine, find out about the NGA's Youth Garden Grants program, search the extensive NGA Library for data on a wide variety of gardening topics, and even check out a "seed swaps" section.

You can also subscribe to a free e-mail newsletter or ask gardening questions. Actually, 16,000 questions have already been answered, so there's a good chance that the information you want is already there waiting for you.

For bulb gardening the Bulb Lady, Debbie Van Bourgondien, has a colorful, useful website at www.dutchbulbs.com. Here you'll find timely articles about virtually every kind of bulb. If the feature articles don't answer your question, just ask the Bulb Lady. She'll e-mail you a personalized answer. Also on this site is a colorful catalog of bulb plants assembled in one convenient place.

When you need just the right tool or piece of equipment and can't find it locally, check the Gardener's Supply Company site, www.gardeners.com. This is devoted primarily to fine merchandise found in its printed catalogs and periodically includes bargains not in the catalog. In addition, you can search an extensive Q&A library of gardening information.

Better yet, you can wander the Biblical plant world too and a 625-acre Biblical Garden Preserve in the Holy Land at www.neot-kedumim.org.il. Neot Kedumim is a labor of love, recreating an extensive preserve with the flowers, herbs, vegetables, fruits, trees, and also some of the fauna mentioned in the Scriptures. You'll find my list of Favorite Biblical Garden websites below, combined with a list of Biblical Gardens to visit with mailing addresses and other contact information.

For organic gardening, and that's immensely popular today, look up Organic Gardening at www.organicgarden.com. They're America's experts and advice is plentiful.

As you surf the web, you'll find hundreds of articles, info pages, and other material you may wish to save. Most sites allow you to download and print out these pages of helpful ideas, tips, and advice that are of interest to you. Be aware, of course, that sites and e-mails change as do people and contacts at these companies. Fortunately, new gardening companies are sprouting and growing, especially those with specialty plants and products. A time-honored Biblical saying guides us to new growing ground: "Seek and ye shall find."

MY FAVORITE GARDEN CATALOGS

Appalachian Gardens, PO Box 82, Waynesboro, PA 17268 (Rare trees/shrubs)

Bluestone Perennials, 7211 Middle Ridge Road, Madison, OH 44057 (Nice variety)

Burgess Seed/Plant Co., 904 Four Seasons Road, Bloomington, IL 61701 (Bulbs, seeds)

Burpee, 300 Park Avenue, Warminster, PA 18974 (Seeds/ bulbs/ plants)

Clyde Robin Seed Co., PO Box 2366, Castro Valley, CA 94546 (Wildflowers)

The Cook's Garden, P.O. Box 535, Londonderry, VT 05148 (Special salad/veggie varieties)

Crystal Palace Perennials, PO Box 154, St. John, IN 46373 (Water garden plants)

Drip Rite Irrigation, 4235 Pacific Street. Suite H, Rocklin, CA 95747 (Irrigation supplies)

Dutch Gardens, PO Box 200, Adelphia, NJ 07710 (Dutch bulbs)

ALLAN SWENSON'S FAVORITE GARDEN WEBSITES

Antique Flowers	www.selectseeds.com
Burpee	www.burpee.com
Cook's Garden	www.cooksgarden.com
Clyde Robin Wildflower Seeds	www.clyderobin.com
Drip Rite Irrigation	www.dripirr.com
Gardens Alive	www.gardensalive.com
Gardener's Supply Company	www.gardeners.com
Garden to the Kitchen	www.gardentokitchen.com
Harris Seeds	www.harrisseeds.com
Johnny's Selected Seeds	www.johnnyseeds.com
Lilypons Water Gardens	www.lilypons.com
Neot Kedumim, Israel	www.neot-kedumim.org.il
Nichols Garden Nursery	www.pacificharbor.com/nichols
Mailorder Gardening Assn.	www.mailordergardening.com
Mountain Valley Growers, Inc.	www.mountainvalleygrowers.com
National Gardening Assn.	www.garden.org
Old House Gardens	www.oldhousegardens.com
Organic Gardening	www.organicgardening.com
Park Seeds	www.parkseed.com
Stokes Tropicals	www.stokestropicals.com
Thompson & Morgan	www.thompson-morgan.com
Van Bourgondien Bulbs	www.dutchbulbs.com
White Flower Farm	www.whiteflowerfarm.com
Wildseed Farms, Ltd.	www.wildseedfarms.com

Flowery Branch Seeds, Box 1330, Flowery Branch, GA 30542 (Rare, heirloom, medicinal)

Forest Farm Nursery, 990 Tetherow Road, Williams, OR, 97544 (Good source)

Gardener's Supply Co., 128 Intervale Road, Burlington, VT 05401 (Many gardening supplies)

Gardens Alive, 5100 Schenley Place, Lawrenceburg, IN 47025 (Organic gardening source)

Harris Seeds, 60 Saginaw Drive, Rochester, NY 14692 (Old line seed firm)

J. L. Hudson—Seedsman, Star Route 2, Box 337, La Honda, CA 94020 (Biblical plants)

Ed Hume Seeds, Inc., PO Box 1450, Kent, WA 98035 (Short season varieties, plus)

Johnny's Selected Seeds, RR 1, Box 2580, Albion, ME 04910 (Seeds, wide variety list)

Klehm's Song Sparrow Perennial Farm, 13101 East Rye Road, Avalon, WI 53505 (Specialities)

Lilypons Water Gardens, PO Box 10, Buckeystown, MD 21717 (Great water garden source)

Mantis, 1028 Street Road, Southampton, PA 18966 (Tillers and tools)

Mellinger's, 2310 W. South Range Road, North Lima, OH 44452 (Variety)

Miller Nurseries, 5060 West Lake Road, Canandaigua, NY 14224 (Great berry/fruit tree source)

Nichols Garden Nursery, 1190 N. Pacific Highway NE, Albany, OR 97321 (Many Asian, international)

Northwoods Nursery, 27635 S. Oglesby Road, Canby, OR 97013 (Rare fruits, nuts, others)

One Green World, 28696 S. Cramer Road, Modalla, OR 97038 (Rare international plants)

Park Seed, 1 Parkton Avenue, Greenwood, SC 29647 (Major U.S. seed and plant firm)

Quality Dutch Bulbs, 13 McFadden Road, Easton PA 18045 (Many bulb flowers)

Roris Gardens, 8195 Bradshaw Road, Sacramento, CA 95829 (Iris specialists)

Royal River Roses, PO Box 370, Yarmouth, ME 04096 (Rare, hardy, oldtime roses)

Seeds of Perfection, PO Box 86, Station A, Toronto, Ont., Canada M9C 4V2 (Unique)

Select Seeds Antique Flowers, 180 Stickney Road, Union, CT 06076 (Heirloom seeds, plants)

Stokes Seeds, Box 548, Buffalo, NY 15240 (Many varieties)

Stokes Tropicals, PO Box 9868, New Iberia, LA 70562 (Exotic tropical plants)

Van Bourgondien, PO Box 1000, Babylon, NY 11702 (Major Dutch bulb specialist)

Vessey's Seeds Ltd,. PO Box 9000, Calais, ME 04619 (US/Canadian varieties)

Wayside Gardens, Hodges, SC 29695 (Major plant source)

White Flower Farm, PO Box 50, Litchfield, CT 06759. (Specialists in rare bulb flowers)

Wildseed Farms, 525 Wildflower Hills, Fredericksburg, TX 78624 (Wildflower specialists)

Hundreds of free garden catalogs are available if you know where to ask. A handy *Garden Catalog Guide* contains descriptions and contact information for 125 garden catalog companies and garden magazine publishers. Also included are smart shopper tips and a glossary of gardening terms. To receive a copy, send a $2 check or money order to Mailorder Gardening Association, Dept. SC-AS, PO Box 2129, Columbia, MD 21045.

This guide also lists dozens of garden firms with toll-free telephone numbers, advice hotlines, e-mail addresses, and websites to browse colorful pages at your leisure.

MAIL-ORDER SHOPPING TIPS

Gardening continues to be America's most popular leisure activity. According to a recent Gallup survey, approximately 64 percent of America's 100.4 million households participate in lawn and garden activities. Many depend on mail-order catalogs to deliver the quality gardening products they demand. More than $2 billion is spent annually on gardening supplies by mail order. There's good reason for this.

Not only do catalogs offer more gardening choices than typical local garden stores, but shopping by mail is convenient, fun, and virtually risk-free, considering the plant guarantees most firms offer today. Mail-order gardening is deeply rooted in America. Today the selection of companies and their specialty seeds, plants, and products is wider than ever. Here are some points that can help as you shop by mail and when you receive your orders. The Mailorder Gardening Association likes happy customers, so it has provided these worthwhile guidelines.

Buy selections appropriate for your climate and garden setting. Most catalogs will provide zone ranges for each plant. Some will provide a map showing the zones so that you can determine which zone your home is in. If not, check with the company or your local County Extension Service to determine your zone.

Be ready when your order arrives. This means having the bed or ground tilled and ready for planting The quicker you can get live plants into the ground, the better. We're all busy, but plants need immediate attention on arrival. Most mail-order firms ship plants and bulbs to arrive at the appropriate planting time for your region. You should plant seeds when appropriate for your area.

If you need help or have questions, call and ask. Most mail-order companies have customer service people to answer questions. They can be very helpful and are there to serve you. You can access the websites I have provided and these provide contact telephone numbers and usually e-mail addresses too. Some have most-asked questions and answers already posted to help gardeners solve basic problems, as well as trickier ones.

Order by mail early to avoid sold-out notices. Because seeds and planting stock are produced at least a season in advance, quantities often are limited. For that reason it pays to order early, especially when a sale catalog arrives.

It is a common practice for many mail-order firms to substitute a similar item for one that is sold out. Usually there is a place on the order to check if you prefer no substitutions. That may be truer for those of us seeking special Biblical plants or closest related plants. Be sure to specify clearly on the order form that you do not want substitutions made, unless you are more flexible to accept them. Keep a record of your purchases: names, item numbers, prices, and dates to facilitate communications between you and the company regarding an order.

These firms receive thousands of orders every week during peak season, so they need your help to track your order with specific details if a problem occurs. That information helps the company call up your order on its computers so it can quickly answer your questions. On the next page is a sample to copy and use for your mail ordering.

Locate and understand the company's guarantee policy. Most firms offer outstanding guarantees. Usually there is a cutoff date by which a company must be notified of problems or plant failures. The date is generally set late enough to allow you sufficient time to plant your order and observe growth. Be sure to inform the company of plant failures or problems before this date.

Order appropriate fertilizers and supplies with your plants. That way you'll have them on hand when planting. Using the right starter fertilizer and soil amendments from the beginning ensures that plants have the best chance of taking root well and prospering.

Finally, as you review the myriad of marvelous mail-order catalogs available, look for tips and ideas to help you make the best choices. More gardening catalogs also offer useful tips for growing plants, especially to guide you to success with the more difficult ones. Take advantage of the catalogs' expertise to create a healthy, beautiful, and bountiful garden.

MY PLANT MAIL ORDER

Company Name_____ Date of Order_____

Catalog/Brochure # _____ Ordering Phone # _____

Customer Account # _____

Customer Service Phone # _____

Items Ordered:

Item #	Name	Quantity	Price

Shipping and Handling Charge $_____

Total $_____

Discounts or coupons applied _____

Check # _____ Card Used_____

Date Order Rec'd. _____ Invoice # _____

Guarantee Cutoff Date _____

Condition of Plants and Notes

_____

_____

_____

_____

_____

WHAT TO DO WHEN YOUR
MAIL ORDER ARRIVES

You may think this part seems silly, but it is one area many people overlook in their busy lives, to the detriment of the plants they order and their garden success. My objective is to provide you with as many worthwhile bits of gardening wisdom that I have collected and learned over the years.

1. Open the package and make sure your order is complete and correct. Check your order form to make sure all of the seed varieties, plants, supplies, and other items have arrived. There may be a note stating that other packages will arrive later. Open any plants wrapped in plastic to allow air circulation. Don't be alarmed if you see dried foliage on dormant plants. Because they are dormant, dried foliage from the prior year is natural and will soon be replaced by new growth.

2. Look for instructions for temporary handling. These are usually included as part of the planting instructions. Handling will vary from plant to plant, but you'll find a few general rules for different types of planting stock here, courtesy of my mail-order garden friendly advisors.

3. Plant as soon as possible. Until you do, give the planting stock proper care to maintain its viability. Follow the instructions provided with your shipment. The company's experts have experience from years in the business and they want you to be well pleased when you follow their useful advice.

Follow these rules for temporary handling of seeds. Simply store your seeds in a cool, dry place until it is time to plant outdoors, or start them indoors in pots and trays if you wish.

For bulbs, corms, rhizomes, and tubers, follow these tips. Keep bulbs such as tulips and daffodils in a cool, dry spot with good air circulation until they can be planted. Remove them from the packaging and spread them in a single layer. Keep them dry and avoid temperature extremes. The ideal time to plant such bulbs is in the fall, after temperatures have permanently cooled and before the onset of winter freezes that harden the ground.

You don't want to plant them too early, nor wait until the ground is unworkable. Remember, always plant bulbs pointy end up.

In contrast to these bulbs, lilies and other bulbs that are not winter hardy should be stored in the dark in a closed box and lightly sprayed with water occasionally to keep them moist until planted. Rhizomes, such as irises, can be kept in their packaging material as long as the shipping carton is open. Store them where they receive some light, but are not exposed to direct sun or wind.

Bare-root stock is standard for mail-order shipping. Many perennials, shrubs, and even trees are shipped bare root, without soil, in a dormant state. Until planting, keep the roots in their protective wrapping of plastic, newspaper, burlap, and wood shavings, in which they were shipped. Moisten them frequently and keep them from exposure to direct sun. The greatest danger is excessive drying. Before planting roses, shrubs, and trees, soak their roots in water for a few hours. Carry them to the garden in their water bucket and plant them directly from the water to avoid any drying prior to planting.

If you are unable to plant for a longer period, a week or more, it is advisable to "heel in" bareroot plants. Heeling-in is a form of temporary planting. Dig a V-shaped trench deep enough to hold the roots. Place the plants in a trench so they are sitting at about a 45-degree angle, and cover the roots with an equal mix of builder's sand and peat moss. If soil is workable, you can use it instead of the sand-peat mix. Keep plants well watered, especially if temperatures are warm, until you are able to uncover and plant them where they will grow in your garden or landscape area.

Green plants need special attention. Green plants in nursery pots are in their growth cycle and require the most careful handling Remove them from their packaging, water them, and place them where they receive the proper amount of light according to specific instructions, and the proper range of temperatures, especially if they are tender plants.

Stock in small pots can dry out quickly, so keep close watch on them and plant as soon as possible. For container stock planted at midsummer, cut back tops by one-third to prevent die-back.

You can apply these tips to bulbs, roots, and plants you buy locally, of course, if you are faced with a tight work schedule and can't get into the garden ground immediately.

Too often many gardeners have a variety of ongoing projects that require attention. These tips will help you have more success with them as part of your good growing horizons.

Chapter Twelve

❧

Favorite Biblical Garden
Websites Around the World

Biblical gardens are sprouting, blooming, and growing across America and around the world. Thanks to the Internet and cyberspace, today you can visit some of them with a click of your mouse. Many have glorious photos of flowers and landscapes plus trees, herbs, fruits, and vegetables. Even better, there is a wealth of valuable information, about the plants they grow, links to other Biblical gardens and plant information, and beautiful color photos to enjoy too. Each reflects the dedication and love that the founders, directors, and those who tend them faithfully have invested and willingly share. With a salute also to the webmasters who have carefully planted these virtual cyberspace gardens and cultivated them on their website pages, my gratitude and compliments.

Sometimes useful new research and gardening tools seem to just sprout. Biblical garden friends sent a relatively new one along so I have added it here as we prepare to go to press. The goal of gardenprojectonline.com is to activate peoples' faith using the garden as a "teaching tool." Steve Hallowell started the project because he felt that more people could be attracted to a relationship with Christ if they could relate it to something familiar like gardening. As one visitor put it, "You really try to give faith a little visual substance to go on." Much like Jesus used parables, gardenprojectonline.com is trying to find a "way in" to activate people's faith. My goal is to get people up "out of the pews" and into an everyday relationship with God. Gardening is a wonderful vehicle to soften people's soil, and is a living symbol of God's miracles.

That seems to fit nicely with my longtime living and writing theme of *Let's Grow Together* which also is the foundation of this book, helping share Biblical gardening ideas, tips, and advice gathered from many gardeners with readers everywhere.

NEW BIBLICAL GARDEN INFORMATION WELCOMED

I always welcome information about other Biblical gardens and gardening ideas that deserve to be celebrated and saluted. Please feel welcome to e-mail me about such gardens with the names of people who tend them at aswenson@gwi.net so that they may receive the recognition they deserve when I revise the next edition of this book. Here are those marvelous Biblical garden websites that you can use to expand your Biblical gardening horizons.

Temple Beth Shalom of Sun City, Arizona
    www.goodnet.com/~tbsaz/body_index.html
Paradise Valley United Methodist Church, Paradise Valley, Arizona
    www.pvumc.org/about/biblicalgarden.html
Magnolia Plantation and Gardens, Charleston, South Carolina
    www.magnoliaplantation.com/gardens/index.html
Warsaw Biblical Garden, Warsaw, Indiana
    www.warsawbiblicalgarden.org
First Congregational Church, Fair Haven, Vermont
    www.sover.net/~hkfamily
Cathedral of St. John the Divine, New York, New York
    www.stjohndivine.org
Rodef Shalom Biblical Gardens, Pittsburgh, Pennsylvania

www.rodefshalom.org/Garden/initial.htm
Temple Sinai Biblical Gardens, Newport News, Virginia
www.ujcvp.org/temple_sinai/bulletin04-00
St. John's Episcopal Church, Norman, Oklahoma
www.episcopalnorman.org/outreach.htm#garden
First Presbyterian Church, Nashville, Tennessee
www.klis.com/scove/041400.htm
First Presbyterian Church, Gainesville, Florida
http://www.gnv.fdt.net/~hpc/garden/
Biblical Garden References Shirley Sidell, Founder
www.BiblicalGardens.com
Bible Garden Links for Adults
www.suite101.com/articles
First Biblical Resources U.S.A.
www.brusa.org/biblegardenpage.htm
Brinkman Country Gardens, Canada
http://www.granite.sentex.net/~lwr/brickman.html
Neot Kedumim, Lod, Israel
http://www.neot-kedumim.org.il
Yad-Hasmona Biblical Gardens, Israel
http://www.yad8.com/bg/biblical.htm
Biblical Garden, Elgin, Scotland
http://www.dufus.com/Duffus+2000biblical_garden/

A helpful website for study of the Bible, comparisons, and research of many topics, is www.biblestudytools.com. I've found it
useful in a variety of ways.

FAVORITE BIBLICAL PLANT SOURCES

Finding appropriate Biblical flowers has often been a difficult
task for those starting out with a plan, a goal, a vision, but only a
few bulb, seed, and plant sources. Talking with many accomplished Biblical gardeners and scanning more than 60 mail-
order catalogs, here's a combined list of sources that have proved
reliable, reasonable, and also have a range of varieties from

which to select those that are best suited for Biblical gardens. Also, because these established mail-order firms are growing plants themselves, it is worth checking their websites periodically for new flowers and also useful growing tips.

Bluestone Perennials, 7211 Middle Ridge Road, Madison, OH 44057. Reasonably priced and good selection. Tel: 800-852-5243 and 440-428-7535. Sarah Boonstra is very helpful. Website: www.bluestoneperennials.com.

Dutch Gardens, P.O. Box 2037, Lakewood, NJ 08701. A longtime supplier of quality flower bulb, anemones, crocus, cyclamen, iris, lilies, tulips, and other fascinating flowering wonders too. Tel: 800-818-3861, and website: www.dutchgardens.com. Joop Visser is the Senior Field Manager.

Forest Farm Nursery, 990 Tetherow Road, Williams, OR, 97544. A new recommendation by several different Biblical gardeners. Write for its catalog.

Johnny's Selected Seeds, Foss Hill Road, RR 1, Box 2580, Albion, ME 04910. A reliable supplier of flowers, herbs, and vegetables focused on organic production methods. Tel: 207-437-4301, and website: www.johnnyseeds.com.

J. L. Hudson—Seedsman, Star Route 2, Box 337, La Honda, CA 94020. Suggested by several Biblical gardeners. An ethnobotanical catalog is available.

Meadowsweet Herb Farm, North Shrewsbury, VT 05738. Specializes in herbs, but some of those also bear lovely flowers, as sage does. Tel: 802-492-3565.

Mellinger's, 2310 W. South Range Road, North Lima, OH 44452. Has a variety of flowers from seeds and bulbs. Tel: 1-800-321-7444, and website: www.mellingers.com.

Miller Nurseries, 5060 West Lake Road, Canandaigua, NY. Reasonably priced, hardy fruit, nut, and foliage trees and shrubs. I've bought from them for years with good results. Tel: 800-836-9630.

Mountain Valley Growers, 38325 Pepperweed Road, Squaw Valley, CA 93675. Many different biblical varieties by mail order.

Tel: 559-338-2775, and website: www.mountainvalleygrowers.com/bibleherbgarden.

Old House Gardens, 536 Third Street, Ann Arbor, MI 48103. Heirloom bulbs with rare old hyacinths, narcissi, tulips, and others. Unique, informative catalog. Tel: 734-995-1486 and website: www.oldhousegardens.com.

Park Seed, 1 Parkton Avenue, Greenwood, SC 29647. A large mail-order firm with a wide selection of flowers and seeds of less available varieties. Website: www.parkseed.com.

Quality Dutch Bulbs, 13 McFadden Road, Easton, PA 18045. Many varieties and types of flower bulbs.

Richters, Goodwood, Ontario, Canada, is recommended by several gardeners. The website: www.richters.com.

Select Seeds, Antique Flowers, 180 Stickney Hill Road, Union, CT 06076. Delightful, hard-to-find flowers with informative catalog. Marilyn Barlow has assembled an excellent array of old-time, heirloom flowers, Biblical ones, and many others worth growing for fun.

Seeds of Distinction, P.O. Box 86, Station A, Toronto, Canada M9C 4V2. Many hardy flowers and Biblical types. Tel: 416-255-3060. Website has special capitalization: www.SEEDSof DISTINCTION.com.

Territorial Seed Company, P.O. Box 158, Cottage Grove, OR 97424. A treasure of information and good seeds at fair prices. Tel: 541-942-9547, and website is also helpful at: www.territorial-seed.com.

Van Bourgondien, 245 Route 109, P.O. Box 1000, Babylon, NY 11702. A major, longtime flower bulb firm with many Biblical plants. Has useful information at website and offers planting tips and advice. Debbie Van Bourgondien is known nationally as The Bulb Lady, a knowledgeable expert. Tel: 800-622-9997, and website: www.dutchbulbs.com.

Wayside Gardens, 1 Garden Lane, Hodges, SC 29695. Has quality plants of harder-to-find species. Introducing exceptional look-alike Hellebore roses. Tel: 800-845-1124, and website: www.waysidegardens.com.

White Flower Farm, P.O. Box 50, Litchfield, CT 06759. Specialists in rare bulb flowers and it bought Daffodil Mart with its many rare, heirloom type flowers.

As we grow into the new millennium, here are my ten Great Growing Commandments I've put together over the 40+ years that I've been writing about gardening, nature, the environment, and outdoors. Follow them and you'll enjoy deserved Great Growing for years to come.

1. Understand what growing conditions different plants need and provide them.

2. Pull small weeds regularly so they don't steal nutrients and water from good plants.

3. Buy best seed varieties to grow the most beautiful flowers and prolific vegetables.

4. Give vegetables a balanced fertilizer diet regularly so they grow most productively.

5. Recycle organic matter into compost to improve your soil and enrich your garden.

6. Plant an extra row of vegetables for the hungry to donate to your local food pantry.

7. Always read and heed complete label directions before applying any pesticides.

8. Try growing wild for a carefree, natural growing experience with wildflowers.

9. Lend a hand to children to help them to dig in and grow better in many ways.

10. Count your blessings every day as you create a more beautiful, tasteful, healthier world, wherever you live.

Appendix A

❧

Scriptural References to Flowers of the Bible

In the beginning we are given the first revealing references to Biblical plants. As we then read the Bible, we can find other references to specific flowers. I've included all my favorite and major Scriptural references to flowers with the flowers themselves and then combined them in this composite list for convenience.

> "Let the earth bring forth grass, the herb yielding seed, and the fruit tree yielding fruit after his kind, whose seed is in itself, upon the earth: and it was so. And the earth brought forth grass, and the herb yielding seed after his kind, and the tree yielding fruit, whose seed was in itself, after his kind: and God saw that it was good."—King James Version

SCRIPTURES RELATING TO LILIES

"I am a rose of Sharon, a lily of the valleys. As a lily among brambles, so is my love among maidens."—Song of Solomon 2:1–2

Flowers With Traceable Scriptural References

Flower	Chapter/verse	Latin Name
Anemone	Matthew 6:28	*Anemone coronary*
Saffron Crocus	Isaiah 35:1,	*Crocus sativum*
	Song of Solomon 4:13	
Cyclamen	Song of Solomon	*Cyclamen persicum*
Daffodil	Matthew 6:29	*Narcissus sp.* or *tazetta*
Trumpet Narcissus	Isaiah 35:1	
Hyacinth	Song of Solomon 6: 2-3	*Hyacinthus orientalis*
Iris, Blue Flag	Hosea 14:5	*Iris versicolor*
Iris, Yellow Flag	Hosea 14:5	*Iris pseudacorus*
Lily, Madonna	Song of Songs 2:1–2	*Lilium candidum*
Narcissis, Daffodil	Matthew 6:30	*Narcissus tazetta L.*
Star-of-Bethlehem	II Kings 6:25	*Ornithogalum umbellatum*
Tulip, Red Sharon	Song of Solomon 2:12	*Tulipa sp.*

"I will be as the dew to Israel; he shall blossom as the lily, he shall strike root as the poplar."—Hosea 14:5

"The desert shall rejoice and blossom; like the crocus it shall blossom abundantly, and rejoice with joy and singing."—Isaiah 35:1–2

"His cheeks are as a bed of spices, as sweet flowers: his lips like lilies, dropping sweet smelling myrrh."—Song of Solomon 5:13

ATTRACTIVE ANEMONES— *ANEMONE CORONARIA*

"And why take ye thought for raiment? Consider the lilies of the field, how they grow; they toil not, neither do they spin: And yet I say unto you, That Solomon in all his glory was not arrayed like one of these."— Matthew 6:28–29

"Consider the lilies how they grow: they toil not, they spin not; and yet I say unto you, that Solomon in all his glory was not arrayed like one of these. If then God so clothe the grass, which is today in the field, and tomorrow is cast into the oven; how much more will he clothe you, O ye of little faith?"—Luke 12:27–28

CROCUS—*CROCUS SATIVUS*

"Spikenard and saffron; calamus and cinnamon, with all trees of frankincense; myrrh and aloes, with all the chief spices . . ."—Song of Solomon 4:14

"Your shoots are an orchard of pomegranates with all choicest fruits, henna with nard, nard and saffron, calamus and cinnamon, with all trees of frankincense, myrrh and aloes, with all chief spices."—Song of Solomon 4:13–24

"The desert shall rejoice and blossom; like the crocus it shall blossom abundantly, and rejoice with joy and singing."—Isaiah 35:1–2

CYCLAMEN—*CYCLAMEN PERSICUM*

"Consider the lilies how they grow: they toil not, they spin not; and yet I say unto you, that Solomon in all his galory was not arrayed like one of these."—Luke 12:27

HYACINTHS—*HYACINTHUS ORIENTALIS*

"My beloved is gone down into his garden, to the beds of spices, to feed in the gardens, and to gather lilies. I am my beloved's, and my beloved is mine: he feedeth among the lilies. Thou art beautiful, O my love, as Tirzah, comely as Jerusalem . . ."—Song of Solomon 6:2–4

IRIS, BLUE FLAG—*IRIS VERSICOLOR,* AND *IRIS ASTROPUPUREA* IRIS, YELLOW FLAG—*IRIS PSEUDACORUS* AND LOOK-ALIKE OPTIONS

"I will be as the dew until Israel: he shall grow as the lily and cast forth his roots as Lebanon."

NARCISSUS—*NARCISSUS THIAZIDE* (DAFFODILS)

"The wilderness and the solitary place shall be glad for them; and the desert shall rejoice, and blossom as the rose,"—Isaiah 35:1

"I am the rose of Charon, and the lily of the valleys. As the lily among thorns, so is my love among the daughters."—Song of Solomon 2:1–2

STAR OF BETHLEHEM—DOVE'S DUNG, *ORNITHOGALUM UMBELLATUM*

"And there was a great famine in Samaria: and, behold, they besieged it, until an ass's head was sold for fourscore pieces of silver, and the fourth part of a cab of dove's dung for five pieces of silver."—II Kings 6:25

TULIP—*TULIPA MONTANA*

"I am the rose of Charon and the lily of the valleys."—Song of Solomon 2:1

"The flowers appear on the earth, the time of singing has come and the voice of the turtledove is heard in our land."—Song of Solomon 2:12

Appendix B

Comparisons of Biblical/ Scriptural Quotations From Different Translations

For centuries people have debated which flowers are meant by which Scriptures. Part of the confusion is based on the different translations and versions of the Bible. To better demonstrate how this situation may have confused gardeners as they tried to determine which flowers to grow in their Biblical gardens, I've compiled comparisons here. You can also do your own homework if you wish to expand your Biblical horizons. A wonderful help, which I discovered thanks to my devout cousin, Amy Tudor Dugdale Luciano, is on the Internet at www.biblestudytools.net. That saved me thumbing through a dozen different Bibles for this list of comparisons.

LUKE 12:27–28—RE: LILIES AND/OR OTHER FLOWERS OF THE BIBLE

King James Version, Luke 12:27–28: "Consider the lilies how they grow: they toil not, they spin not; and yet I say unto you, that Solomon in all his glory was not arrayed like one of these. If then God so clothe the grass, which is today in the field, and tomorrow is cast into the oven; how much more will he clothe you, O ye of little faith?"

Douay Version, Luke 12:27–28: "Consider the lilies, how they grow: they labour not, neither do they spin. But I say to you, not even Solomon in all his glory was clothed like one of these. Now, if God clothe in this manner the grass that is to-day in the field and to-morrow is cast into the oven: how much more you, O ye of little faith."

New Revised Standard, Luke 12:27–28: "Consider the lilies, how they grow: they neither toil nor spin; yet I tell you, even Solomon in all his glory was not clothed like one of these. But if God so clothes the grass of the field, which is alive today and tomorrow is thrown into the oven, how much more will he clothe you—you of little faith!"

New International Version, Luke 12:27–28: "Consider how the lilies grow. They do not labor or spin. Yet I tell you, not even Solomon in all his splendor was dressed like one of these. If that is how God clothes the grass of the field, which is here today, and tomorrow is thrown into the fire, how much more will he clothe you, O you of little faith!"

The New English Bible, Luke 12:27–28: "Think of the lilies: they neither spin or weave; yet I tell you, even Solomon in all his splendour was not attired like one of these. But if that is how God clothes the grass, which is growing in the field today, and tomorrow is thrown on the stove, how much more will he clothe you! How little faith you have!"

The Bible—An American Translation—Edgar J. Good-speed, Luke 12:27–28: "See how the lilies grow. They do not toil or spin, but I tell you, even Solomon in all his splendor was never dressed like one of them. But if God so dresses the wild grass, which is alive today, and is thrown into the furnace tomorrow, how much more surely will he clothe you, who have so little faith?"

Good News for Modern Man, Luke 12:27–30: "Which one of you can live a few years more by worrying about it? And why worry about clothes. Look how the wild flowers grow: they do not work or make clothes for themselves. But I tell you that not even Solomon, as rich as he was, had clothes as beautiful as one of these flowers. It is God who clothes the wild grass—grass that is here today, gone tomorrow, burned up in the oven. Will he not be all the more sure to clothe you? How little is your faith."

MATTHEW 6:28–30—RE: LILIES AND OTHER FLOWERS OF THE SCRIPTURES

King James Version, Matthew 6:28–30: "And why take ye thought for raiment? Consider the lilies of the field, how they grow; they toil not, neither do they spin: And yet I say unto you, that even Solomon in all his glory was not arrayed like one of these. Wherefore, if God so clothe the grass of the field, which to day is, and to morrow is cast into the oven, shall he not much more clothe you, O ye of little faith!"

Douay-Rheims Bible, Matthew 6:28–30: "And for raiment why are you solicitous? Consider the lilies of the field, how they grow: they labour not, neither do they spin. But I say to you, that not even Solomon in all his glory was arrayed as one of these. And if the grass of the field, which is today, and to morrow is cast into the oven, God doth so clothe: how much more you, O ye of little faith?"

New Revised Standard Version, Matthew 6:28–30: "Consider the lilies how they grow: they neither toil nor spin; yet I tell you, even Solomon in all his glory was not clothed like one of these. But if God so clothes the grass of the field, which is alive today and tomorrow is thrown into the oven, how much more will he clothe you—you of little faith!"

New International Version, Matthew 6:28–30: "Who of you by worrying can add a single hour to his life? And why do you worry about clothes? See how the lilies of the field grow. They do not labor or spin. Yet I tell you that not even Solomon in all his splendor was dressed like one of these. If that is how God clothes the grass of the field, which is here today and tomorrow is thrown into the fire, will he not much more clothe you, O you of little faith?"

The New English Bible, Matthew 6:28–30: "Is there a man of you who by anxious thought can add a foot to his height? And why be anxious about clothes? Consider how the lilies grow in the fields; they do not work, they do not spin. And yet I tell you, even Solomon in all his splendour was not attired like one of these. But if that is how God clothes the grass in the fields, which is there today, and tomorrow is thrown into the stove, will he not all the more clothe you?"

The Bible—An American Translation—Edgar J. Goodspeed, Matthew 6:28–30: "But which of you with all his worry can add a single hour to his life? Why should you worry about clothing? See how the wild flowers grow. They do not toil or spin, and yet I tell you, even Solomon in all his splendor was never dressed like one of them. But if God so beautifully dresses the wild grass, which is alive today and is thrown into the furnace tomorrow, will he not much more surely clothe you, you who have so little faith?"

Good News for Modern Man, Matthew 6:28–30: "And why worry about clothes? Look how the wild flowers grow: they do not work or make clothes for themselves. But I tell you that not even Solomon, as rich as he was, had clothes as beautiful as one of these flowers. It is God who clothes the wild grass—grass that is here today, gone tomorrow, burned up in the oven. Will He not be all the more sure to clothe you? How little is your faith!"

II KINGS 6:25—RE: STAR OF BETHLEHEM

New International Version, II Kings 6:25: "There was a great famine in the city; the siege lasted so long that a donkey's head sold for eighty shekels of silver, and a quarter of a cab of seed pods for five shekels."

The New English Bible, II Kings 6:25: "The city was near starvation, and they besieged it so closely that a donkey's head was sold for eighty shekels of silver, and a quarter of a kab of locust beans for five shekels."

King James Version, II Kings 6:25: "And there was a great famine in Samaria: and behold, they besieged it, until an ass's head was sold for fourscore pieces of silver and the fourth part of a cab of dove's dung for five pieces of silver."

The Bible—An American Translation—Edgar J. Goodspeed, II Kings 6:25: "Consequently there was a great famine in Samaria, and there the besiegers continued until an ass's head was sold for eighty shekels of silver, and a pint of dove's dung for five shekels of silver."

The Holy Scriptures—According to the Masoretic Text, II Kings 6:25: "And there was a great famine in Samaria; and behold, they besieged it, until an ass's head was sold for fourscore pieces of silver, and the fourth part of a kab of dove's dung for five pieces of silver."

SONG OF SONGS 1:14—RE: HYACINTH, NARCISSUS, AND TULIP

New International Version, Song of Songs 1:14: "My lover is to me a cluster of henna blossoms from the vineyards of En Gedi."

The New English Bible, Song of Songs 1:14: "My beloved is for me a cluster of henna-blossom for the vineyards of En-gedi."

King James Version, Song of Songs 1:14: "My beloved is unto me as a cluster of camphire in the vineyards of En-gedi."

The Bible—An American Translation—Edgar J. Goodspeed, Song of Songs 1:14: "A cluster of henna is my beloved to me, from the gardens of Engedi."

The Holy Scriptures—According to the Masoretic Text, Song of Songs 1:14: "My beloved is unto me as a cluster of henna in the vineyards of En-gedi."

SONG OF SONGS 2:1–2—RE: HYACINTH, NARCISSUS, AND TULIP

King James Version, Song of Songs 2:1–2: "I am the rose of Sharon, and the lily of the valleys. As the lily among thorns, so is my love among the daughters."

New International Version, Song of Songs 2:1–2: "I am a rose of Sharon, a lily of the valleys. Like a lily among thorns is my darling among the maidens."

The New English Bible, Song of Songs 2:1-2: "I am an asphodel in Sharon, a lily growing in the valley. No, a lily among thorns is my dearest among girls."

The Bible—An American Translation—Edgar J. Goodspeed, Song of Songs 2:1–2: "I am a saffron of the plain, a hyacinth of the valleys. Like a hyacinth among thistles so is my loved one among the maidens."

The Holy Scriptures—According to the Masoretic Text, Song of Songs 2:1–2: "I am a rose of Sharon, a lily of the valleys. As a lily among thorns, so is my love among the daughters."

SONG OF SONGS 4:13–14—RE: CROCUS

King James Version, Song of Songs 4:13–14: "Thy plants are an orchard of pomegranates, with pleasant fruits; camphire, with spikenard, spikenard and saffron; calamus and cinnamon, with all trees of frankincense; myrrh and aloes, with all the chief spices."

New International Version, Song of Songs 4:13–14: "Your plants are an orchard of pomegranates with choice fruits, with henna and nard, nard and saffron, calamus and cinnamon, with every kind of incense tree, with myrrh and aloes and all the finest spices."

The New English Bible, Song of Songs 4:13–14: "Your two cheeks are an orchard of pomegranates, an orchard full of rare fruits: spikenard and saffron, sweet-cane and cinnamon, with ever incense-bearing tree, myrrh and aloes with all the choicest spices."

The Bible—An American Translation—Edgar J. Goodspeed, Song of Songs 4:13–14: "Your products are a park of pomegranates, together with choice fruits, henna with nard and saffron, calamus and cinnamon, together will all kinds of frankincense woods, myrrh and aloes, together with all the finest perfumes."

The Holy Scriptures—According to the Masoretic Text, Song of Songs 4:13–14: "Thy shoots are a park of pomegranates, with precious fruits; henna with spikenard plants, spikenard and saffron, calamus and cinnamon, with all trees of frankincense, myrrh and aloes, with all the chief spices."

SONG OF SONGS 6:2–4—RE: HYACINTH

New International Version, Song of Songs 6:2–4: "My lover has gone down to his garden, to the beds of spices, to browse in the gardens and to gather lilies. I am my lover's and my lover is mine; he browses among the lilies."

The New English Bible, Song of Songs 6:2–4: "My beloved has gone down to his garden, to the beds where balsam grows, to delight in the garden and to pick the lilies. I am my beloved's and my beloved is mine, he who delights in the lilies."

The Bible—An American Translation—Edgar J. Goodspeed, Song of Songs 6:2–4: "My beloved has gone down to his garden, to the beds of spices, to pasture his flock in the gardens and gather hyacinths. I belong to my beloved and my beloved to me, who pastures his flock among the hyacinths."

The Holy Scriptures—According to the Masoretic Text, Song of Songs 6:2–4: "My beloved is gone down to his garden, to the beds of spices, to feed in the gardens and to gather lilies. I am my beloved's and my beloved is mine, that feedeth among the lilies."

SONG OF SONGS 5:13—RE: LILY

King James Version, Song of Songs 5:13: "His cheeks are as a bed of spices, as sweet flowers; his lips like lilies, dropping sweet smelling myrrh."

New International Version, Song of Songs 5:13: "His cheeks are like beds of spice, yielding perfume. His lips are like lilies, dripping with myrrh."

The New English Bible, Song of Songs 5:13: "His cheeks are like beds of spices or chests full of perfumes; his lips are lilies, and drop liquid myrrh."

The Bible—An American Translation—Edgar J. Goodspeed, Song of Songs 5:13: "His cheeks are like beds of spices, exhaling perfumes. His lips are hyacinths, dropping flowing myrrh."

The Holy Scriptures—According to the Masoretic Text, Song of Songs 5:13: "His cheeks are as a bed of spices, as banks of sweet herbs, his lips are as lilies, dropping with flowing myrrh."

ISAIAH 35:1–2—RE: CROCUS

King James Version, Isaiah 35:1–2: "The wilderness and the solaria place shall be glad for them; and the desert shall rejoice, and blossom as the rose. It shall blossom abundantly and rejoice even with joy and singing.

New International Version, Isaiah 35:1–2: "The desert and the parched land will be glad; the wilderness will rejoice and blossom. Like the crocus, it will burn into bloom; it will rejoice greatly and shout for joy."

The New English Bible, Isaiah 35:1–2: "Let the wilderness and the thirsty land be glad, let the desert rejoice and burst into flower. Let it flower with fields of asphodel, let it rejoice and shout for joy."

The Bible—An American Translation—Edgar J. Goodspeed, Isaiah 35:1–2: "The wilderness and the parched land shall be glad, and the desert shall rejoice and blossom; like the crocus shall it blossom abundantly; it shall rejoice with joy and singing."

The Holy Scriptures—According to the Masoretic Text, Isaiah 35:1–2: "The wilderness and the parched land shall be glad; and the desert shall rejoice, and blossom as the rose. It shall blossom abundantly, and rejoice, even with joy and singing."

HOSEA 14:5—RE: IRIS

King James Version, Hosea 14:5: "I will be as the dew unto Israel: he shall grow as the lily, and cast forth his roots as Lebanon."

New International Version, Hosea 14:5: "I will be like the dew to Israel; he will blossom like a lily. Like a cedar of Lebanon, he will send down his roots; his young shoots will grow."

The New English Bible, Hosea 14:5: "I will be as dew to Israel, that he may flower like the lily, strike root like the poplar and put out fresh shoots."

The Bible—An American Translation—Edgar J. Goodspeed, Hosea 14:5: "I will be like the dew to Israel, so that he will blossom like the lily, and his roots will spread like the poplar; his tendrils will spread out."

The Holy Scriptures—According to the Masoretic Text, Hosea 14:5: "I will be as the dew unto Israel, he shall blossom as the lily and cast forth his roots as Lebanon. His branches shall spread and his beauty shall be as the olive tree, and his fragrance as Lebanon."

OTHER BIBLICAL FLOWERS WITH SCRIPTURAL COMPARISONS—RE: COMMON POPPY, *PAPAVER RHOEAS:* DOG CHAMOMILE, *ANTHEMIS SP.*

Isaiah 40:6, 8: "All flesh is grass, and all its beauty is like the flower of the field. . . . The grass withers, the flower fades; but the word of our God will stand for ever."

I Peter 1:24–25: "All flesh is like grass and all its glory like the flower of the grass. The grass withers, and the flower falls, but the word of the Lord abides forever."

James 1:9–10: "Let the lowly brother boast in his exaltation and the rich in his humiliation, because like the flower of the grass he will pass away."

Bibliography

I gratefully thank and salute all who have carefully researched, and dug deeply in the Scriptures, botany, natural history, and other topics to create their books. My Biblical plant book library is 4 feet long and still growing—pun intended. One of my goals in writing this book is to help promote the planting and cultivation of more Biblical flower gardens across America and around the entire world. Not just flowers, but also Biblical gardens with an entire range of plants of the Scriptures and Holy Land. The more you read and learn about this fascinating field of gardening, the easier it will be for you and your friends to become Biblical gardeners and apostles for Biblical gardens in your area.

My favorite reference books are at the top of this list, mainly because they are available today from bookstores, Internet sources, and some from out-of-print book specialists. Other reference books are mostly out-of-print but can sometimes be found through state library search systems, horticultural libraries, or rare book dealers. Good reading, researching and gardening!

MOST AVAILABLE BIBLICAL PLANT RESOURCE BOOKS

Hareuveni, Nogah, *Nature in Our Biblical Heritage*, 1980, 146 pages, illustrated.

Hareuveni, Nogah, *Tree and Shrub in Our Biblical Heritage*, 1984, 146 pages, illustrated.

Hareuveni, Nogah, *Desert and Shepherd in Our Biblical Heritage*, 1991, 160 pages, illustrated.

Hareuveni, Nogah, *Ecology in the Bible*, 1974, 52 pages, 62 color photos.

Hepper, F. Nigel, *Baker Encyclopedia of Bible Plants, Flowers, Trees, Fruits, Vegetables, Ecology*, 1992, 190 pages, well illustrated.

Hepper, F. Nigel, *Planting a Bible Garden*, 1987 and 1997, 92 pages, illustrated.

Hudson-Knapp, Rev. M., *Plants in a Biblical Garden, Prayer Guide to the Children's Vegetable and Herb Garden*, A Biblical Garden Database via *hkfamily@sover.net*.

King, E. A., *Bible Plants for American Gardens*, 1941 and 1975, revised edition.

Moldenke, H. N., *Plants of the Bible*, 1940, 135 pages.

Moldenke, H. N., and A. L. E., *Plants of the Bible*, 1952, 328 pages, reprint, 1986.

Swenson, Allan A., *Your Biblical Garden*, Doubleday edition, 1981, 220 pages.

Swenson, Allan A., *Plants of the Bible and How to Grow Them*, 1995, 220 pages.

Zohary, Michael, *Plants of the Bible*, 1982, 220 pages.

OTHER BIBLICAL PLANT RESOURCE BOOKS

Alon, Azaria, *The Natural History of the Land of the Bible*, 1969, 276 pages.

Balfour, J. H., *The Plants of the Bible; Trees and Shrubs*, 1857. Enlarged edition, 250 pages, 1885.

Cotes, R. A., *Bible Flowers*, 1904, 288 pages.

Crowfoot, G. M. and L. Baldensperger, *From Cedar to Hyssop: A Study in the Folklore of Plants in Palestine*, 1904, 204 pages.

Natural History of the Bible, 1824, 462 pages.

Hepper. F. Nigel, *Bible Plants at Kew*, 1981 and 1985.

Hiller, N., J. D. Johnson, & D. J. Wiseman, *The Illustrated Bible Dictionary*, 1980, 3 vols.

James, Wilma, *Gardening with Biblical Plants*, 1983.

Paterson, J. & K., *Consider the Lilies, Plants of the Bible*, 92 pages, 1986.

Reicke, B. *The New Testament Era: the World of the Bible*, 1969.

Temple, A. A., Flowers and Trees of Palestine, 1907, 184 pages.

Untermeyer, L., *Plants of the Bible*, 1970, 26 pages.

Vester, Bertha Spafford, *Flowers of the Holy Land*, 17 original watercolors, 1962, 64 pages.

Zohary, Michael, *Flora of the Bible*, Interpreters Dictionary of the Bible, 1962.

Zohary, Michael, *The Plant Life in Palestine*, 1962.

Also check out the American Horticultural Society's *A-Z Encyclopedia of Garden Plants*. Almost all Biblical plants are included with growing zones, soil conditions, and other cultivation tips.

BIBLES FOR BIBLICAL SCRIPTURE COMPARISONS AND PLANT IDENTIFICATIONS

Bible, English. Holy Bible, Authorized King James Version, 1967.

Bible, English, Revised Standard Version translated from the original languages as the version set forth in 1611, revised 1946–1952 and 1971.

Bible, Authorized King James Version, translated out of the original tongues and with the former translations diligently compared and revised, Judson Press, 1942.

Moffatt, J. The Old Testament: a new translation. Volume I, Genesis to Esther, 571 pages, 1924. Volume II, Job to Malachi, 482 pages, 1925.

Bible, Good News for Modern Man, The New Testament in Today's English Version, American Bible Society, 1966.

Bible, The New English Bible, Standard Edition, Oxford University and Cambridge University Press, 1970.

Bible, Life Application Study Bible, New International Version, Tyndale House Publishers, and Zondervan Publishing House, 1997.

The Holy Scriptures, according to the Masoretic text, Menorah Press, 1957, 1960, 1973.

Moffatt, J., The New Testament: a new translation together with the Authorized Version. Parallel edition with introduction, 676 pages, 1922.

Phillips, J. B., The New Testament in Modern English, 1960.

Goodspeed, Edgar J., Popular Edition, An American Translation, 1935.

Wright, C. H., The Holy Bible—Containing the Old and New Testaments translated out of the original tongues. Edited by C. H. Wright, 268 pages, 1895.

Oxford University Press, The Holy Bible, containing Old and New Testaments, 1056 pages containing selected helps to study of the Bible.

Bible, Various Translations, Versions at Bible Study Tools website: www.biblestudytools.com.

Young's Analytical Concordance, by Robert Young, LL.D., is a valuable resource. English Bible translators were not botanists. In attempting to identify actual Biblical plants, it is better to search by their Hebrew or Greek names. This concordance organizes plants and other words based on their Greek and Hebrew names.

Cruden's Concordance, by Alexander Cruden, M.A., remains a valuable guide to the Scriptures of the Old and New Testaments of the King James Version. Since local libraries may not have or be able to easily obtain the reference books you wish to read, you may find it helpful to visit or contact one of the many horticultural libraries in the United States. The libraries specialize in books and other writings concerned with the field of horticulture. Theological seminaries and other religious institutions, from colleges to divinity schools, are also potential reference sources, but their number precludes listing in this book.

The Brooklyn Botanic Garden has two excellent research libraries and its catalog is available online at http://www.bbg.org/ lib/index.html. Its Gardener's Resource Center, accessible from that same site, will answer questions about any kind of plant or plant process. A list of my favorite garden Information websites is included elsewhere in this book and provides a wealth of links to horticultural experts at major companies in the United States and also around the world. Tap their talents for great growing knowledge.

Index

NOTE: *italic page numbers* indicate charts and illustrations